ECOSYSTEM SERVICES

FOR SUSTAINABILITY

宝库山精选：可持续性生态系统服务

Digital editions

Ecosystem Services for Sustainability is available through most major ebook and database services (please check with them for pricing). Special print/digital bundle pricing is also available in cooperation with Credo Reference; contact Berkshire Publishing (info@berkshirepublishing.com) for details.

For information, contact:
Berkshire Publishing Group LLC
122 Castle Street
Great Barrington, Massachusetts 01230-1506 USA
www.berkshirepublishing.com
Printed in the United States of America

Library of Congress Cataloging-in-Publication Data

Ecosystem services for sustainability / Ray C. Anderson, general editor.
    p. cm.—(Berkshire essentials)
  Includes bibliographical references and index.
    ISBN 978-1-61472-966-2 (pbk. : alk. paper)—ISBN 978-1-61472-967-9
(ebook) 1. Ecosystem management. 2. Biodiversity conservation.
3. Human ecology. I. Anderson, Ray C.
  QH75.E3235 2013
  333.72—dc23   2013027647

BERKSHIRE 宝库山

Essentials

ECOSYSTEM SERVICES FOR SUSTAINABILITY

宝库山精选：可持续性生态系统服务

BERKSHIRE
A global point of reference

About *Ecosystem Services for Sustainability*

Ecosystem *Services for Sustainability*, a Berkshire Essential, covers the many benefits provided by things we too often take for granted: the aquifers and glaciers that provide our drinking water, the forests and oceans that store excess carbon dioxide, the microscopic organisms that enrich the soil in which we grow our food, and the insects that play a crucial role in maintaining the global food supply. A team of well-known authors, writing for the nonexpert reader, looks at how nature provides for humankind, and how humankind can, in return, protect valuable and limited natural resources. The authors also explore practical ways of managing ecosystem services and integrating them into our global economy.

THE **BERKSHIRE** *Essentials* SERIES

Berkshire Sustainability Essentials, distilled from the *Berkshire Encyclopedia of Sustainability*, take a global approach to environmental law, energy, business strategies and management, industrial ecology, and religion, among other topics.

- Religion and Sustainability
- Business Strategies and Management for Sustainability
- Energy Resources and Sustainability
- Energy Industries and Sustainability
- Environmental Law and Sustainability
- Finance and Investment for Sustainability
- Industrial Ecology

Distilled for
the classroom
from Berkshire's
award-winning
encyclopedias

BERKSHIRE ESSENTIALS **from the** *Berkshire Encyclopedia of China* **and the** *Berkshire Encyclopedia of World History, 2nd Edition* **also available.**

Contents

About Berkshire Essentials

For more than a decade, Berkshire Publishing has collaborated with a worldwide network of scholars and editors to produce award-winning academic resources on popular subjects for a discerning audience. The "Berkshire Essentials" series are collections of concentrated content, inspired by requests from teachers, curriculum planners, and professors who praise the encyclopedic approach of Berkshire's reference works, but who still crave single volumes for course use.

Each Essentials series draws from Berkshire publications on a big topic—world history, Chinese studies, and environmental sustainability, for instance—to provide thematic volumes that can be purchased alone, in any combination, or as a set. Teachers will find the insightful articles indispensable for stimulating classroom discussion or independent study. Students, professionals, and general readers all will find the articles invaluable when exploring a line of research or an abiding interest.

These affordable books are available in paperback as well as ebook formats for convenient reading on mobile devices.

Editors, Editorial Advisory Board, and Production Staff

Editors

The following people served as Editors for the sources of these articles (all from the *Berkshire Encyclopedia of Sustainability*):

From Volume 1, *The Spirit of Sustainability:* Willis Jenkins, General Editor, *Yale University;* Whitney Bauman, *Florida International University*

From Volume 2, *The Business of Sustainability:* Chris Laszlo, General Editor, *Case Western Reserve University;* Karen Christensen, *Berkshire Publishing Group;* Daniel S. Fogel, *Wake Forest University;* Gernot Wagner, *Environmental Defense Fund;* Peter Whitehouse, *Case Western Reserve University*

From Volume 3, *The Law and Politics of Sustainability:* Klaus Bosselmann, *University of Auckland,* Daniel S. Fogel, *Wake Forest University,* J.B. Ruhl, *Vanderbilt University Law School*

From Volume 4, *Natural Resources and Sustainability:* Daniel E. Vasey, General Editor, *Divine Word College,* Sarah E. Fredericks, *University of North Texas,* Lei Shen, *Chinese Academy of Sciences,* Shirley Thompson, *University of Manitoba*

From Volume 5, *Ecosystem Management and Sustainability:* Robin Kundis Craig, *University of Utah,* John Copeland Nagle, *University of Notre Dame,* Bruce Pardy, *Queen's University,* Oswald J. Schmitz, *Yale University,* William K. Smith, *Wake Forest University*

From Volume 6, *Measurements, Indicators, and Research Methods for Sustainability:* Ian Spellerberg, General Editor, *Lincoln University,* Daniel S. Fogel, *Wake Forest University,* Sarah E. Fredericks, *University of North Texas,* Lisa M. Butler Harrington, *Kansas State University*

Editorial Advisory Board

Ray Anderson, *Interface, Inc.;* Lester Brown, *Earth Policy Institute;* Robert Costanza, *University of Vermont;* Luis Gomez-Echeverri, *United Nations Development Programme;* John Elkington, *SustainAbility;* Daniel Kammen, *University of California, Berkeley;* Ashok Khosla, *International Union for Conservation of Nature;* and Christine Loh, *Civic Exchange, Hong Kong*

Production Staff

Project Coordinator
Bill Siever

Copy Editors
Linda Aspen-Baxter
Mary Bagg
Kathy Brock

Carolyn Haley
Barbara Resch
Elma Sanders
Chris Yurko

Editorial Assistants
Echo Bergquist

David Gagne
Ellie Johnston

Designer
Anna Myers

Introduction: What are Ecosystem Services?

The benefits people obtain from nature are known as ecosystem services. The concept has come to play important roles in both research and policy since the 1970s. Ecosystem services are being integrated more explicitly into multilateral environmental agreements, national accounting frameworks, corporate strategy, and public policy. Accounting for many different ecosystem services at a sufficiently large scale to promote sustainability, however, remains a future challenge.

People depend on nature for their livelihoods, health, and welfare. The benefits that people get from nature include clean water for drinking, food and recreation from fishing, and wood for building houses and furniture. At the same time, people affect nature in ways that limit its ability to provide these benefits. For instance, forests help to regulate climate by capturing and storing carbon, but each year landholders reduce forests' ability to provide this service by clearing thousands of hectares of forests in the tropics. Clearing this land releases up to 20 percent of all human emissions of carbon dioxide—a greenhouse gas that contributes to climate change. This is just one example of the ways we are transforming nature and altering the benefits—or ecosystem services—that nature provides to people.

What Are Ecosystem Services?

The most common definition of ecosystem services comes from the United Nations Millennium Ecosystem Assessment (MA): "the benefits people obtain from ecosystems" (MA 2005). Ecosystem services are also referred to as *environmental goods and services* and *nature's benefits*. The services flow from the functions and processes of ecosystems, including the species that make them up

(Daily 1997). The MA (2005, 57) identifies four categories of ecosystem services:

1. *provisioning services* that deliver goods such as food, water, timber, and fiber
2. *regulating services* that stabilize climate, moderate risk of flooding and disease, and protect or enhance water quality
3. *cultural services* that offer recreational, aesthetic, educational, and spiritual experiences
4. *supporting services* that underpin the other services, such as photosynthesis and nutrient cycling

Alternative definitions and classifications have been proposed for specific contexts, such as landscape management, environmental accounting, and policy development (Boyd and Bhanzaf 2006; De Groot, Wilson, and Roelof 2002; Fisher, Turner, and Morling 2009; Wallace 2007). In 2010, the Economics of Ecosystems and Biodiversity (TEEB), an international initiative led by the United Nations Environment Programme (UNEP), proposed a definition that differentiates between the services provided by ecosystems and the benefits that humans receive from them: "the direct and indirect contributions of ecosystems to human well-being" (Kumar 2010, 19). The TEEB classification for ecosystem services redefines *supporting services* as ecosystem *processes* and includes a new category of *habitat services*, which provide nurseries for hunted or fished species, and preserve future options by protecting genetic diversity.

History

The understanding that people rely on nature for their well-being dates to antiquity. Some of the earliest known texts on this topic describe the loss of ecosystem services and the impact of that loss on society. Chief among these is a description in *Critias*, one of the famous dialogues of the Greek philosopher Plato:

> What now remains of the formerly rich land is like the skeleton of a sick man with all the fat and soft earth having wasted away and only the bare framework remaining. . . . Once the land was enriched by yearly rains, which were not lost, as they are now, by flowing from the bare land into the sea. The soil was deep, it absorbed and kept the water . . . and the water that soaked into the hills fed springs and running streams everywhere. Now the abandoned shrines at spots where formerly there were springs attest that our description of the land is true. (Daily 1997, 5–6)

Many scholars trace modern concern about ecosystem services to George Perkins Marsh, a nineteenth-century lawyer, politician, and scholar. Marsh's 1864 book *Man and Nature* describes a range of services and the consequences of their loss. In the first half of the twentieth century, prominent environmental writers, including Henry Fairfield Osborn Jr., William Vogt, and Aldo Leopold, wrote about the value of ecosystems and wildlife for human welfare. In addition to nature's value for people, Leopold also espoused a *land ethic* that places a value on the existence of nature itself, without regard to the ways humans use it.

Environmental health became an important issue during the 1960s and 1970s, sparking the first ecological economics research. In 1968, Stanford University ecologist Paul Ehrlich published *The Population Bomb*, which describes human disruption of ecosystems, the costs for society, and possible solutions. In 1970, the Study of Critical Environmental Problems, a group of scientists meeting together at Williams College in Massachusetts, presented the term *environmental services* for the first time, with examples such as fisheries, climate regulation, and flood control. Since then, *ecosystem services* has become the most common term in the scientific literature for benefits that derive from nature.

By the 1980s, research and debate centered on two questions: how much ecosystem function and services depend on biodiversity, and how to measure and value ecosystem services. In 1997, two groups of ecologists and economists synthesized scientific information about ecosystem services and their value (Costanza et al. 1997; Daily 1997). The Millennium Ecosystem Assessment (MA) that began in 2001 involved 1,360 researchers in a four-year global study that evaluated the state of ecosystems and the services they provide. The MA reported that, out of twenty-four ecosystem services tracked over the previous fifty years, fifteen services had seriously declined, four had shown some improvement, and five were generally stable but under threat in some parts of the world (MA 2005). The assessment also revealed that some provisioning services, such as food, had improved at the expense of regulating, supporting, and cultural services. A number of smaller assessments of ecosystem services were undertaken as part of, and following, the MA. One study found that of the ecosystem services delivered by eight broad aquatic and terrestrial habitat types in the United Kingdom and their constituent biodiversity, about 30 percent were declining and others were reduced or degraded (UK National Ecosystem Assessment 2011). A separate international study of the Economics of Ecosystems and Biodiversity (TEEB) assessed the economic benefits of ecosystems and biodiversity and the costs of ecosystem degradation and biodiversity loss (Kumar 2010). As of 2011, several countries, including Brazil and India, were conducting national-level TEEB studies.

Emerging Policies and Programs

Spurred in part by international assessments like the MA, various organizations instituted new policies and programs for ecosystem services. In 2005, the United Nations (UN) began to consider establishing a new UN authority that would review ecosystem service research and disseminate conclusions relevant to policy making. As a result, the Intergovernmental Platform for Biodiversity and Ecosystem Services (IPBES) was established in 2010. Also in 2010, the United Nations Framework Convention on Climate Change (UNFCCC) adopted an international framework for reducing greenhouse gas emissions from deforestation and forest degradation (REDD), which provides a way for industrialized countries to offset their emissions by purchasing credits from developing countries that store additional carbon in their forests. The World Bank has developed a partnership for wealth accounting and the valuation of ecosystem services (WAVES) to encourage and enable countries to incorporate nature's value into national accounting frameworks and indicators, such as gross domestic product. In addition, the Convention on Biological Diversity and the Ramsar Convention on Wetlands have explicitly incorporated ecosystem services and ecosystem-based approaches in their principles.

At national and state levels, too, ecosystem service policies and markets are in place. In the United States, the Clean Water Act (1974) protects wetlands and bodies of water to avoid loss of hydrological, cultural, and habitat services. The act provides for mitigating the loss of wetlands and their functions. One provision of the law is that developers who build on wetlands are required to restore or protect an equivalent or greater amount of wetland area to offset the loss of a wetland's fisheries, recreational opportunities, water purification, and erosion control services, among others. In Australia, too, several state laws require mitigation of damage to ecosystems for the habitat services they provide. Brazil's Forest Law (1965) requires landowners in the Amazon to maintain 80 percent of their landholdings under forest to preserve the benefits from intact forests.

Civil Society

Various civil society programs have also emerged since the Millennium Ecosystem Assessment, including widely cited work at the World Resources Institute (WRI), Forest Trends, and the Natural Capital Project. In 2008 the WRI released the guide *Ecosystem Services: A Guide for Decision Makers*, which provides practical guidance on policies that sustain natural capital (Ranganathan et al. 2008). The WRI has also developed the Corporate Ecosystem Services Review (ESR) to help companies identify the ecosystem services they affect and depend on (Hanson et al. 2008). Forest Trends aims to expand the value of nature to society through the creation of markets for ecosystem services. Among other initiatives, Forest Trends has developed a clearinghouse for ecosystem service projects, an ecosystem service project incubator, a program for marine ecosystem services, and a voluntary biodiversity offset framework. The Natural Capital Project is an academic–civil society partnership that develops science and tools to measure, map, and value ecosystem services (see Kareiva et al. 2011); apply these tools with government, business, and civil society partners around the world; and spread the science, tools, and lessons from those efforts worldwide.

Tools

New diagnostic tools for monitoring, measuring, and valuing ecosystem services are constantly being developed. By the end of 2011, more than a dozen ecosystem service assessment tools were in use. Some tools focus on the provision, use, value, and trade-offs of multiple ecosystem services under different resource management scenarios. Two of these are ARtificial Intelligence for Ecosystem Services (ARIES) and Integrated Valuation of Ecosystem Services and Tradeoffs (InVEST). Both ARIES and InVEST use maps to assess the spatial distribution of ecosystem services, as does the Natural Assets Information System (NAIS). Other tools, such as the Wildlife Habitat Benefits Estimation Toolkit and WRI's coral reef valuation package, estimate the value or amount of ecosystem services without spatial representation. Additional tools assess the benefits from a single ecosystem service (for example, carbon sequestration and storage calculators) or consider changes in ecosystem services in a particular context (for example, benchmarking tools for policies and practices in the mining industry).

Payments for Ecosystem Services

One of the areas of greatest growth since about 2000 is payments for ecosystem services (PES), in which users compensate the providers of services for maintaining or enhancing them (Gómez-Baggethun et al. 2010; Wunder, Engel, and Pagiola 2008). In the developing world, two types of PES mechanisms are prominent: payments for watershed services and payments for climate regulation through the REDD framework. Payments for watershed services, also known as water funds, are a way for downstream water users to pay upstream landholders for the delivery of water services, such as purification and erosion control. One of the first PES programs in Costa Rica is described in a later section, Ecosystem Services in Practice.

Valuation Methods

Economic valuation involves assigning a monetary value to nature's benefits. Existing market prices often do not reflect ecosystem service values, and special valuation methods based on similar or hypothetical market situations are required. A 2004 white paper published by the World Bank clarifies the aims and uses of economic valuation, outlining four principle objectives: assessing the value of the total flow of benefits from ecosystems, determining the net benefit of an intervention that alters ecosystem conditions, determining how the costs and benefits of ecosystem conservation are distributed, and identifying beneficiaries to ascertain potential funding sources for conservation (Pagiola, von Ritter, and Bishop 2004).

The analytical approach for economic valuation of ecosystem services must be shaped to meet the specific objective. The framework often used to value these services is total economic value (TEV), which includes the value of direct and indirect use of nature by people, values that are independent of human uses, and option value, or the benefits of preserving an ecosystem for future use. Since the value of nature is not easily characterized—or fully captured—in monetary terms, many studies quantify ecosystem services values in terms of impacts on human health and nutrition, livelihood benefits, and cultural significance. Others simply measure ecosystem services in biophysical, rather than monetary, terms, such as tons of carbon sequestered.

Controversy

The concept of ecosystem services has generated some controversies. One of the most recurrent has its roots in the distinction first made by Aldo Leopold between utilitarian values—those benefits the natural world provides for people—and the intrinsic value of nature—the right of ecosystems and species to exist, regardless of whether humans use or appreciate them. Some scientists have raised concerns that the emerging focus on valuing ecosystem services could shift efforts and funding away from biodiversity and species protection for its own sake, which could increase the rate at which species are lost. Biodiversity is not considered an ecosystem service in many ecosystem service classifications; some ecologists are, however, attempting to identify the ways biodiversity contributes to human well-being and the roles biodiversity plays in critical ecosystem processes that are necessary to provide ecosystem services (for example, see Mace, Norris, and Fitter forthcoming 2012), and they are also studying where ecosystem service provision and biodiversity priorities overlap (Naidoo et al. 2008).

Critics are concerned about valuation as a tool and market-based mechanisms, such as payment for ecosystem services, that turn ecosystem services into commodities or "put a price tag on nature" (Gómez-Baggethun et al. 2010). Some researchers contend that nature is priceless and that market programs do not put a high enough price on them to ensure conservation. Others suggest that market mechanisms might have unintended adverse environmental and social consequences, because consumers sometimes act in unexpected ways when they are given external incentives. In addition, these researchers say that the distribution of financial or other benefits is likely to magnify existing economic disparities.

Ecosystem Services in Practice

Ecosystem services are increasingly being considered in decision making as policy makers take into account the value of services and how actions affect those values; establish innovative market-based mechanisms that ensure service values are reflected in market transactions; implement policy, organizational, and institutional reform; and develop tools that help people to do all of this quickly and easily. Following are several illustrative examples.

Costa Rica

By 1986, forest cover as a share of total land area in Costa Rica had fallen to 32 percent from 63 percent in 1960. This dramatic loss of forest resources prompted the Costa Rican government to pursue a new set of forest conservation and restoration policies, including market-based mechanisms. The passage of Forest Law 7575 in 1996 laid the groundwork for a PES program to pay landowners to maintain and restore forest resources. The program is focused on four ecosystem services generated by forests: watershed protection, biodiversity, landscape beauty, and climate regulation. Program payments to landowners vary by activity. The most common activity is forest protection; landowners agree to forgo use of their forests, transferring their use-rights to the government in exchange for a fixed payment per hectare disbursed over five years. Landowners are also paid by the hectare for reforestation activities or are paid for each tree planted in an agroforestry system, in which crops are interplanted with trees (Fondo Nacional de Financiamiento Forestal n.d.).

China

Following devastating droughts and floods in 1997 and 1998, China instituted a series of conservation programs to reduce damage from extreme weather. One of these

programs, the Sloping Land Conversion Program (SLCP), also known as the Grain to Green Program, is exceptional in its longevity and geographic expanse. Established in 1999, the SLCP restores erosion control and flood mitigation services in twenty-five provinces through grain and cash subsidies to farmers who convert agricultural fields on steep slopes to forests and grasslands. Initial studies have found that the SLCP has increased key ecosystem services while also having a positive effect on household income (Li et al. 2011). Following the SLCP, China has developed Ecosystem Conservation Function Areas (EFCAs), which are newly established zones identified for conservation because of their high levels of biodiversity and ecosystem services, including sediment retention and carbon storage and sequestration. When development is complete, these EFCAs are projected to cover 25 percent of China's landmass. Land use master plans at the provincial and county level will steer development activities away from these areas, mandating low to no infrastructure development in these zones.

Belize

In 1998 and again in 2011, the WRI released a global *Reefs at Risk* report that mapped and analyzed threats to the world's coral reefs in order to visualize where reefs could be lost. In a complementary series of *Coastal Capital* reports, the WRI produced economic valuations of the reefs and mangroves in the Caribbean to raise awareness of the benefits that people get from these ecosystems and build support for policies that promote their sustainable management. In the case of Belize, the study prompted the government to impose a number of fishing restrictions to protect its coastal resources, including size limits for Nassau groupers caught, a ban on spearfishing within marine protected areas, and a mandate that all fish fillets must be brought to landing sites with a skin patch to enable species identification. Moreover, after the container ship *Westerhaven* ran aground on a reef in January 2009, the Belizean government worked with civil society partners to calculate compensatory ecosystem-related damages that were used in a subsequent court case.

Wild Bee Pollination

In 2011, the agribusiness Syngenta assessed the value of wild bee pollination to blueberry farms in Michigan and the added value created by providing foraging habitat for native bees. The purpose of the study was to show that conserving bee populations gave a positive return on investment. The Syngenta valuation was a pilot test for the World Business Council for Sustainable Development's (WBCSD) *Guide to Corporate Ecosystem Valuation*, which is designed to improve companies' understanding of the benefits and value of ecosystem services. The guide explores tools and methods for ecosystem valuation to manage risks and opportunities related to ecosystem services. The Syngenta study determined that Michigan blueberry farmers received $12 million annually from wild bee pollination of their crops. The company has since launched Operation Pollinator to support conservation programs that growers can integrate into their farms (WBCSD and IUCN 2011).

Challenges and Future Directions

The ongoing challenge with ecosystem services is to build on the many new tools and approaches. Many decisions made by individuals, communities, corporations, and governments still do not reflect the value of nature's benefits to people. There are critical gaps in both the scientific basis of ecosystem services and policy and finance mechanisms.

The relationships between ecosystem services and biodiversity, human well-being, and poverty remain unclear. Many studies do not address multiple ecosystem services and their interactions or the consequences that changes in ecosystem services in one place have on distant places (Seppelt et al. 2011). In addition, few systematic studies reveal the effects of different policy instruments on ecosystem services and the people that provide and benefit from them. New global research programs, however, are taking up these challenges.

Ecosystem service programs and policies are often piecemeal and poorly coordinated. In many cases, they are based on unproven assumptions or sparse information (Carpenter et al. 2009). In addition, disproportionately few programs and policies are focused on dryland, grassland, subterranean, or marine ecosystems. In a 2009 study, the Bridgespan Group observed that 73 percent of the ecosystem service projects they looked at focused on forests and wetlands (Searle and Cox 2009). Moreover, many existing policies and programs address one or two ecosystem services, rather than multiple services. Last, relatively few nations have adopted ecosystem service policies, although the number is growing. An important future challenge is to take multiple ecosystem services into account at a large enough scale to ensure that the environment provides the many benefits society needs to prosper.

Amy ROSENTHAL
Natural Capital Project at the World Wildlife Fund

Kimberly LYON
Multilateral Relations at the World Wildlife Fund

Emily McKENZIE
Natural Capital Project at the World Wildlife Fund

See also in the *Berkshire Encyclopedia of Sustainability* Biodiversity; Buffers; Carrying Capacity; Community Ecology; Fisheries Management; Forest Management; Groundwater Management; Human Ecology; Hunting; Microbial Ecosystem Processes; Natural Capital; Nutrient and Biogeochemical Cycling; Ocean Resource Management; Reforestation; Wilderness Areas

FURTHER READING

Boyd, James, & Banzhaf, Spencer. (2006). What are ecosystem services? The need for standardized environmental accounting units. *Ecological Economics, 63*(2–3), 616–626.

Burke, Lauretta; Reytar, Katie; Spalding, Mark; & Perry, Allison. (2011). *Reefs at risk revisited.* Washington, DC: World Resources Institute.

Carpenter, Steven, et al. (2009). Science for managing ecosystem services: Beyond the Millennium Ecosystem Assessment. *PNAS, 106*(5), 1305–1312.

Costanza, Robert, et al. (1997). The value of the world's ecosystem services and natural capital. *Nature, 387*(6630), 253–260.

Daily, Gretchen. (Ed.). (1997). *Nature's services.* Washington, DC: Island Press.

De Groot, Rudolf S.; Wilson, Matthew A.; & Roelof, M. J. Boumans. (2002). A typology for the classification, description and valuation of ecosystem functions, goods and services. *Ecological Economics, 41*(3), 393–408.

Fisher, Brendan; Turner, R. Kerry; & Morling, Paul. (2009). Defining and classifying ecosystem services for decision making. *Ecological Economics, 68,* 643–653.

Fondo Nacional de Financiamento Forestal. (n.d.). Homepage. Retrieved September 20, 2011, from http://www.fonafifo.go.cr/

Gómez-Baggethun, Erik; de Groot, Rudolf; Lomas, Pedro L.; & Montes, Carlos. (2010). The history of ecosystem services in economic theory and practice: From early notions to markets and payment schemes. *Ecological Economics, 69*(6), 1209–1218.

Hanson, Craig; Finisdore, John; Ranganathan, Janet; & Iceland, Charles. (2008). The corporate ecosystem services review: Guidelines for identifying business risks & opportunities arising from ecosystem change. Washington, DC: World Resources Institute.

Kareiva, Peter; Tallis, Heather; Ricketts, Taylor H.; Daily, Gretchen C.; & Polasky, Stephen. (Eds.). (2011). *Natural capital: Theory & practice of mapping ecosystem services.* Oxford, UK: Oxford University Press.

Kumar, Pushpam. (Ed.). (2010). *The economics of ecosystems and biodiversity: Ecological and economic foundations.* London: Earthscan.

Li, Jie; Feldman, Marcus W.; Li, Shuzhuo; & Daily, Gretchen C. (2011). Rural household income and inequality under the Sloping Land Conversion Program in western China. *PNAS, 108*(19), 7721–7726.

Mace, Georgina; Norris, Ken; & Fitter, Alastair H. (forthcoming 2012). Biodiversity and ecosystem services: A multi-layered relationship. *Trends in Ecology and Evolution.*

Millennium Ecosystem Assessment (MA). (2005). *Ecosystems and human well-being: Synthesis.* Washington, DC: Island Press.

Naidoo, Robin, et al. (2008). Global mapping of ecosystem services and conservation priorities. *PNAS, 105*(28), 9495–9500.

Pagiola, Stefano; von Ritter, Konrad; & Bishop, Joshua. (2004). *How much is an ecosystem worth? Assessing the economic value of conservation.* Washington, DC: World Bank.

Ranganathan, Janet, et al. (2008). *Ecosystem services: A guide for decision makers.* Washington, DC: World Resources Institute.

Ruhl, J. B.; Kraft, Steven; & Lant, Christopher. (2007). *The law and policy of ecosystem services.* Washington, DC: Island Press.

Searle, Bob, & Cox, Serita. (2009). *The state of ecosystem services.* Retrieved September 10, 2011, from http://www.moore.org/files/The%20State%20of%20Ecosystem%20Services.pdf

Seppelt, Ralf; Dormann, Carsten F.; Eppink, Florian V.; Lautenback, Sven; & Schmidt, Stefan. (2011). A quantitative review of ecosystem service studies: Approaches, shortcomings and the road ahead. *Journal of Applied Ecology, 48*(3), 630–636.

Tallis, Heather; Goldman, Rebecca; Uhl, Melissa; & Brosi, Berry. (2009). Integrating conservation and development in the field: Implementing ecosystem service projects. *Frontiers in Ecology and the Environment, 7*(1), 12–20.

UK National Ecosystem Assessment. (2011). *The UK National Ecosystem Assessment: Synthesis of the key findings.* Cambridge, UK: UNEP-WCMC.

Wallace, Ken. (2007). Classification of ecosystem services: Problems and solutions. *Biological Conservation, 139*(3–4), 235–246.

World Business Council for Sustainable Development (WBCSD) & International Union for Conservation of Nature (IUCN). (2011). *Guide to corporate ecosystem valuation: A framework for improving corporate decision-making.* Geneva: World Business Council for Sustainable Development.

Wunder, Sven; Engel, Stefanie; & Pagiola, Stefano. (2008). Taking stock: A comparative analysis of payment for environmental services programs in developed and developing countries. *Ecological Economics, 65,* 834–85.

Air Pollution Indicators and Monitoring

As air pollution has become widely recognized as an issue affecting human health and the sustainability of the environment, scientists and policy makers have developed indices to measure and report on levels of pollution. Although they are in widespread use, the indices have not been standardized, and different indices measure different components of pollution.

Anthropogenic air pollution (pollution caused by people) has existed since prehistoric humans' first fires. Even the ancient Greeks and Romans were aware of the problem. As early as 1661, the English author John Evelyn wrote about London's air quality in a pamphlet called *Fumifugium*, stating "inhabitants breathe nothing but an impure and thick mist, . . . corrupting the lungs and disordering the entire habit of their bodies." The Industrial Revolution resulted in further widespread air pollution. Since the 1950s, a variety of legislation has been introduced to reduce emissions, but, at the same time, increased use of motor vehicles has introduced new air pollutants. Outdoor air pollution is now estimated to cause 1.3 million deaths worldwide per year (WHO 2011a).

The public is often confused by the information it receives about pollution concentrations and health effects. To help clarify the situation, several agencies have formulated air quality indices to provide a universal platform to compare and assess the impact of the various pollutants. Air quality monitoring, however, requires a network of sophisticated and high-maintenance equipment. The use of bioindicators—organisms that exhibit different responses to different levels of pollution or by accumulation of a pollutant—has been growing. Biomonitoring is used to complement existing air pollution networks.

Background

Air pollution is the presence of any physical, chemical, or biological agent that modifies the natural characteristics of the atmosphere. According to the US Environmental Protection Agency, air pollution is "the presence in the outdoor atmosphere of one or more air contaminants in sufficient quantities and of such characteristics and duration as to be injurious, or tend to be injurious to human health and welfare, plant or animal life, or to property, or which unreasonably interferes with the enjoyment of life and property or the conduct of business" (US Department of Commerce 1978). Air pollution depends upon many factors, including climate, as well as political, economic, and industrial development. It can be considered on many levels, from local to global. For example, on a continental scale, acid rain in Scandinavia is considered to be the result of pollution from the United Kingdom and western Europe, while the release of greenhouse gases is a global problem.

The composition and relative level of pollutants in the atmosphere differs significantly in urban and rural environments, resulting from a variety of natural sources (such as volcanoes, dust storms, forest fires, and pollens) and anthropogenic activities (such as industrial processes, motor vehicle exhaust, heat and power generating facilities, and combustion). All of these produce pollutants that impair the quality of the atmosphere. The use of biofuels in homes in developing countries not only pollutes the outdoor environment but is also responsible for poor indoor air quality and results in around 2 million deaths annually in the developing world (WHO 2011a).

The Industrial Revolution, technological advancements, and the increased use of motor vehicles have contributed greatly to the deterioration of air quality. At the same time, the impact of air pollution on human health and the environment began to be systematically documented during this same period. Scientists have identified a range of ambient air pollutants and their emission sources, but data relating to their concentrations and impact on human health and ecosystems is available for only a limited number of air pollutants.

Air pollutants exist as aerosols (tiny liquid and solid particles), particulates (solid particles that may be larger in size), and gases. They can enter the air directly (primary pollutants, such as oxides of sulfur, nitrogen, and carbon; organic compounds; particulate matter; and metal oxides) or be formed via chemical reactions in the atmosphere (secondary pollutants formed under the influence of light energy: photochemical oxidants). Along with six criteria pollutants (carbon monoxide [CO], nitrogen dioxide [NO_2], ozone [O_3], lead [Pb], particulate matter [PM], and sulfur dioxide [SO_2]), the United States Environmental Protection Agency (EPA) has identified a large number of air pollutants that are known to cause or may reasonably be believed to cause adverse effects on human health or adverse environmental effects. Initially, 187 specific pollutants and chemical groups were identified as hazardous air pollutants (HAPs), and the list has been modified over time. For example methyl ethyl ketone was removed from the list in 2005, ethylene glycol monobutyl ether in 2004, and caprolactam in 1996 (US EPA 2007).

Due to improved emission control technologies and strict legislation, the levels of ambient air pollutant concentrations in developed countries have fallen considerably, but they remain a serious environmental health problem in a large number of low- and middle-income countries, particularly in megacities (those with more than 10 million people). The World Health Organization (WHO) estimates that a quarter of the world's population is exposed to unhealthy concentrations of air pollutants, and outdoor air pollution is responsible for 1.3 million deaths annually around the world (WHO 2011a).

Among air pollutants, fine particulate matter is of greatest concern due to its association with a variety of acute and chronic illnesses, such as lung cancer and cardiopulmonary diseases. (Because particulate matter is so small, it can easily be inhaled.) According to WHO, fine particulate matter is responsible globally for 9 percent of lung cancer deaths, 5 percent of cardiopulmonary deaths, and about 1 percent of respiratory infection deaths (WHO 2011b). National and international ambient air quality regulations and standards have been introduced to protect human health and ecosystems. The threshold concentrations of various pollutants are set after a detailed review of available scientific evidence of their impacts on human health.

To ensure that the established national or international ambient air quality standards are being met,

continuous monitoring of the various air pollutants is carried out at fixed monitoring stations in different locations. The public, however, typically is unaware of the links between air pollution and health and does not really understand existing air quality information. Air quality indices have been developed as a tool to inform the public of potential health problems related to air quality or pollutant concentrations.

Air Quality Indices

Air quality indices (AQI) are categorical, generally based on numeric ranges, and are computed by different governments and agencies to report the quality of air with reference to health risks. The higher the AQI, the higher the health risk to the population of that region. To calculate the AQI, ambient air concentrations of various pollutants are continuously monitored at nationwide networks of monitoring stations, and these concentrations are transformed into an index. Numerical values are given in ranges with reference to a health risk or level of air pollution, and each range is typically assigned a color code together with a description. The functions and scales used to convert concentrations of various air pollutants to AQI vary widely among countries (Plaia and Ruggieri 2011).

In 1976, the EPA developed the first air quality index, called the Pollutant Standard Index (PSI), based on carbon monoxide, sulfur dioxide, particulate matter with a diameter of 10 micrometers (PM_{10}), ozone, and nitrogen dioxide. In 1999, the Air Quality Index (US EPA 2009 and 2011) replaced the PSI, and the index was extended to include $PM_{2.5}$ (particulate matter with a diameter of 2.5 micrometers) as well as PM_{10}. Despite some limitations (for example, limited applicability to regions other than the United States and no consideration of the cumulative effect of multiple pollutants), the AQI is widely used. Many countries, however, continue to use PSI because they do not monitor $PM_{2.5}$ (Cheng et al. 2007).

The EPA calculates the AQI for five air pollutants regulated by the US Clean Air Act. Twenty-four-hour concentrations of these pollutants are measured and reported in six reference categories and symbolized by different colors. An AQI value of 100 corresponds to the national air quality standard for the pollutant. A value of 50 represents good air quality with little potential to affect public health, while an AQI value over 300 represents hazardous air quality. (See table 1 on page 9.)

A website developed by a number of US agencies provides the public with easy access to national air quality information (AIRNow n.d.). This site provides daily AQI forecasts as well as real-time AQI conditions for more

TABLE 1. EPA Air Quality Index

Air Quality Index Levels of Health Concern	Numerical Value	Meaning
Good	0 to 50	Air quality is considered satisfactory, and air pollution poses little or no risk
Moderate	51 to 100	Air quality is acceptable; however, for some pollutants there may be a moderate health concern for a very small number of people who are unusually sensitive to air pollution.
Unhealthy for Sensitive Groups	101 to 150	Members of sensitive groups may experience health effects. The general public is not likely to be affected.
Unhealthy	151 to 200	Everyone may begin to experience health effects; members of sensitive groups may experience more serious health effects.
Very Unhealthy	201 to 300	Health warnings of emergency conditions. The entire population is more likely to be affected.
Hazardous	301 to 500	Health alert: everyone may experience more serious health effects.

Source: EPA (2009).

This widely distributed chart generally includes color coding from green for good to maroon for hazardous.

than three hundred cities across the United States. It also provides links to more detailed state and local air quality information.

In Europe, methods and systems used to report air quality vary greatly. A recent European Union (EU) project called CITEAIR gives details for the different countries (van der Elshout and Léger 2007).

The United Kingdom has its own index. In its 2012 revision, carbon monoxide was dropped, $PM_{2.5}$ was added, and index bands for PM_{10}, nitrogen dioxide, and ozone were made more stringent (Department for Environment Food and Rural Affairs 2011). The Daily Air Quality Index, as it is now called, reports the level of air pollution in terms of four air pollution bands (low, moderate, high, or very high) and a ten-point index. The air quality index is based on short-term health effects. (See tables 2 and 3, on pages 10 and 11.)

Air quality indices are now in widespread use. There is no consensus, however, either on the scale or the number of bands. Table 4, on page 11, shows that France, for example, uses a scale of 1 to 10, while Australia uses 0 to 200; both have six bands. All air quality indices have a common aim to reduce adverse health effects from short-term increases in air pollution. Some display a bias toward regulatory issues and others to health effects. As shown in table 3, on page 11, the index is often accompanied with health advice. There is no common approach. In some indices the advice varies by pollutant, whereas in others separate advice may be given for non-risk groups and at-risk groups. Without a consistent approach, a comparison of AQI values is difficult and of limited usefulness.

A number of innovative methods have also been used to inform the public about air pollution. These include

Table 2. UK Daily Air Quality Index Bands

Band	Index	Ozone	Nitrogen Dioxide	Sulphur Dioxide	PM2.5 Particles	PM10 Particles
		Running 8 hourly mean	Hourly mean	15 minute mean	24 hour mean	24 hour mean
		μgm^{-3}	μgm^{-3}	μgm^{-3}	μgm^{-3}	μgm^{-3}
LOW						
	1	0–33	0–66	0–88	0–11	0–16
	2	34–65	67–133	89–176	12–23	17–33
	3	66–99	134–199	177–265	24–34	34–49
MODERATE						
	4	100–120	200–267	266–354	35–41	50–58
	5	121–140	268–334	355–442	42–46	59–66
	6	141–159	335–399	443–531	47–52	67–74
HIGH						
	7	160–187	400–467	532–708	53–58	75–83
	8	188–213	468–534	709–886	59–64	84–91
	9	214–239	535–599	887–1063	65–69	92–99
VERY HIGH						
	10	240 or more	600 or more	1064 or more	70 or more	100 or more

Source: COMEAP (2011).

*The concentration of an air pollutant is given in micrograms (one-millionth of a gram) per cubic meter of air, or μgm-3.

Indices in different countries vary widely, but like this UK index, many provide a stepwise decrease in air quality with accompanying health risks.

TABLE 3. Health Advice Relating to the UK Air Quality Index

Air Pollution Banding	Value	Accompanying Health Messages for at-Risk Groups and the General Population	
		At-Risk Individuals	General Population
Low	1–3	Enjoy your usual outdoor activities.	Enjoy your usual outdoor activities.
Moderate	4–6	Adults and children with lung problems (and adults with heart problems) who experience symptoms should consider reducing strenuous physical activity, particularly outdoors.	Enjoy your usual outdoor activities.
High	7–9	Adults and children with lung problems, and adults with heart problems, should reduce strenuous physical exertion, particularly outdoors, and particularly if they experience symptoms. People with asthma may find they need to use their reliever inhaler more often. Older people should also reduce physical exertion.	Anyone experiencing discomfort such as sore eyes, cough, or sore throat should consider reducing activity, particularly outdoors.
Very High	10	Adults and children with lung problems, adults with heart problems, and older people, should avoid strenuous physical activity. People with asthma may find they need to use their reliever inhaler more often.	Reduce physical exertion, particularly outdoors, especially if you experience symptoms such as cough or sore throat.

Source: COMEAP (2011).

TABLE 4. Air Quality Indices in Various Countries and Regions

Country	Index Value	Named Bandings	Pollutants Included
Australia	200	6	$CO, NO_2, O_3, PM_{10}, SO_2$
Belgium	10	10	NO_2, O_3, PM_{10}, SO_2
Canada	100	10	$CO, NO_2, O_3, PM_{10}, SO_2$
European Union (CITEAIR)	100	5	$CO, NO_2, O_3, PM_{10}, SO_2$
France	10	6	NO_2, O_3, PM_{10}, SO_2
Germany	100	6	$CO, NO_2, O_3, PM_{10}, SO_2$
Ireland	100	5	NO_2, O_3, PM_{10}, SO_2
United Kingdom	10	4	$NO_2, O_3, PM_{2.5}, PM_{10}, SO_2$
United States	500	6	$CO, NO_2, O_3, PM_{2.5}, PM_{10}, SO_2$

Source: Adapted from COMEAP (2011).

balloons (in Paris), photo journals (in Beijing), a digital exhibition (in Madrid), a living light (in Seoul), and lasers (in Helsinki). These will continue to evolve, and, no doubt, in the future individuals will be able to access information on their own personal exposure based on knowledge of their daily routine and lifestyle.

Ian COLBECK and Zaheer Ahmad NASIR
University of Essex

See also in the *Berkshire Encyclopedia of Sustainability* Biological Indicators—Species; Carbon Footprint; Environmental Performance Index (EPI); Externality Valuation; Land-Use and Land-Cover Change; Ocean Acidification—Measurement; Reducing Emissions from Deforestation and Forest Degradation (REDD); Regulatory Compliance; Shipping and Freight Indicators; Tree Rings as Environmental Indicators

FURTHER READING

AIRNow. (n.d.). Homepage. Retrieved January 5, 2012, from http://airnow.gov/

Cheng, Wan-Li, et al. (2007) Comparison of the revised air quality index with the PSI and AQI indices. *Science of the Total Environment*, *382*, 191–198.

Committee on the Medical Effects of Air Pollutants (COMEAP). (2011). Review of the UK Air Quality Index. Retrieved January 5, 2012, from http://www.comeap.org.uk/images/stories/Documents/Reports/comeap review of the uk air quality index.pdf

Department for Environment Food and Rural Affairs (Defra). (2011). Notification of changes to the Air Quality Index. Retrieved February 8, 2012, from http://uk-air.defra.gov.uk/news?view=158

Plaia, Antonella, & Ruggieri, Mariantonietta. (2011). Air quality indices: A review. *Reviews in Environmental Science and Biotechnology*, *10*(2), 165–179.

United States Department of Commerce. (1978). Air pollution regulations in state implementation plans: Tennessee (PB-290 291). Retrieved February 8, 2012, from http://www.ntis.gov/search/product.aspx?ABBR=PB290291

United States Environmental Protection Agency (EPA). (2007). Modifications to the 112(b)1 hazardous air pollutants. Retrieved February 8, 2012, from http://www.epa.gov/ttn/atw/pollutants/atwsmod.html

United States Environmental Protection Agency (EPA). (2009). Air Quality Index: A guide to air quality and your health. Retrieved January 5, 2012, from http://www.epa.gov/airnow/aqi_brochure_08-09.pdf

United States Environmental Protection Agency (EPA). (2011). Air Quality Index (AQI): A guide to air quality and your health. Retrieved January 5, 2012, from http://cfpub.epa.gov/airnow/index.cfm?action=aqibasics.aqi

van den Elshout, Sef, & Léger, Karine. (2007, June). Comparing urban air quality across borders. Retrieved January 5, 2012, from http://www.airqualitynow.eu/download/CITEAIR-Comparing_Urban_Air_Quality_across_Borders.pdf

World Health Organization (WHO). (2011a). Air quality and health. Retrieved January 5, 2012, from http://www.who.int/mediacentre/factsheets/fs313/en/index.html

World Health Organization (WHO). (2011b). Global Health Observatory (GHO). Mortality and burden of disease from outdoor air pollution. Retrieved January 5, 2012, from http://www.who.int/gho/phe/outdoor_air_pollution/burden/en/index.html

Aquifers

Aquifers are the largest and most reliable sources of freshwater on the planet, where groundwater is a key component of potentially sustainable development. Well interference, surface-water infringement, water-quality degradation, and land subsidence are possible consequences of groundwater depletion and indicators of unsustainable development. Humankind must recognize the limitations of groundwater development and strive to reconcile demand for the resource against the limits of nature.

An aquifer, generally described as a unit of rock that yields water to a well or spring in usable quantities, is an underground reservoir composed of permeable, water-bearing rock or unconsolidated sediment from which water can be extracted at reasonable cost and in sufficient quantities for drinking water or commercial, industrial, or agricultural purposes. Aquifers are the world's largest and most reliable sources of freshwater. They contain the Earth's groundwater upon which all plants and animals depend for survival. Aquifers allow human populations to be sustained in areas where there are few sources of surface water, but that may also have ample sources of groundwater that are replenished by timely recharge.

Globally speaking, the volume of fresh groundwater is estimated to total about 10.5 million cubic kilometers, or about 1.6 times the annual flow of the Amazon River (Gleick 1996). Groundwater (as opposed to surface water, such as rivers, lakes, and wetlands) is used by about 2 billion people worldwide, making it the most relied-upon natural resource. The annual withdrawal of groundwater is estimated to average between 600 and 700 cubic kilometers, or about 4.5 times the volume of the Dead Sea (Giordano 2009). Despite existing throughout most of the world as a theoretically sustainable water supply, groundwater is one of our most abused natural resources, simply because it is not always appreciated that groundwater is only conditionally renewable. In most areas of the world the main use of groundwater is for irrigation of crops. Use of groundwater for drinking is increasing, and competition is growing among the various users. Vast areas of land have been made productive by the addition of groundwater. Examples are farming in the Sahara, the US High Plains, and other regions with minimal surface water.

Aquifers are fragile. When the rate of groundwater withdrawal exceeds that of aquifer recharge, water levels decline as a preliminary symptom of groundwater exploitation, called *overdrafting*. The nature of overdrafting and resulting decline depends on several hydrogeologic factors, including the imbalance between recharge and discharge in addition to the aquifer's geometry and hydraulic properties. If ignored or improperly managed, overdrafting can eventually remove all recoverable groundwater, rendering the aquifer unsustainable.

Occurrence and Availability of Groundwater

The occurrence and availability of groundwater varies greatly from place to place, depending on the climate and hydrogeologic setting. If an aquifer is adequately replenished and protected from contamination, groundwater can be withdrawn indefinitely. Overexploited or neglected aquifers, on the other hand, can spawn a variety of problems, including potential depletion of the resource, induction of saline or polluted water, and subsidence (i.e., sinking) of land surface.

An important consideration for all aquifers is that locations where surface water replenishes the aquifer must be protected from contamination. In sand and gravel areas there is some possibility of attenuation (i.e., lessening) of contaminants by filtering. Karst, typically occurring in areas of soluble limestone with many faults, fractures, and conduits, is more susceptible to contamination due to less attenuation of contaminants.

Distribution and Examples of Aquifers

Groundwater provides for nearly 80 percent of the drinking water in Europe and Russia, and for an even larger percent in North Africa and the Middle East. Groundwater accounts for 98 percent of Denmark's and 96 percent of Austria's drinking water. Roughly half the drinking water in Asia comes from groundwater, which serves nearly 32 percent of that continent's population. In Canada, 30 percent of the population, or 8.9 million people, rely on groundwater for domestic purposes. An estimated 100 million people rely on groundwater for drinking in the United States, where approximately 33 percent of the public supply and 95 percent of rural domestic supplies depend on groundwater. During 2005, about 23 percent of all freshwater used in the United States came from groundwater sources; 67 percent of that groundwater was used for irrigation, with another 18 percent used for public-supply (mostly drinking water) purposes. (Kenny et al. 2009).

Excluding Antarctica (composed of approximately 98 percent continental ice and 2 percent barren rock), roughly 30 percent of the Earth's continents are underlain by aquifers that contain 30.1 percent of all freshwater not considered surface water or ice. (See figure 1.) About one-half the continental mass contains relatively minor aquifers of unconsolidated material in the shallow subsurface.

One of the world's largest aquifers is the Guarani Aquifer, which covers roughly 1.2 million square kilometers (approximately the size of South Africa) as it stretches from central Brazil to northern Argentina. Another of the world's largest aquifers is the Great Artesian Basin, which supplies water to remote parts of southern Australia. The largest aquifer in the world is the Ogallala Aquifer, which underlies parts of eight states in the central region of the United States. The Ogallala Aquifer supplies about 30 percent of all water used in the United States for irrigation (Weeks et al. 1988). Because water is replenished at exceedingly slow rates (roughly 10 percent of the annual rate of withdrawal), this vast aquifer system

Figure 1. The Distribution of Earth's Water

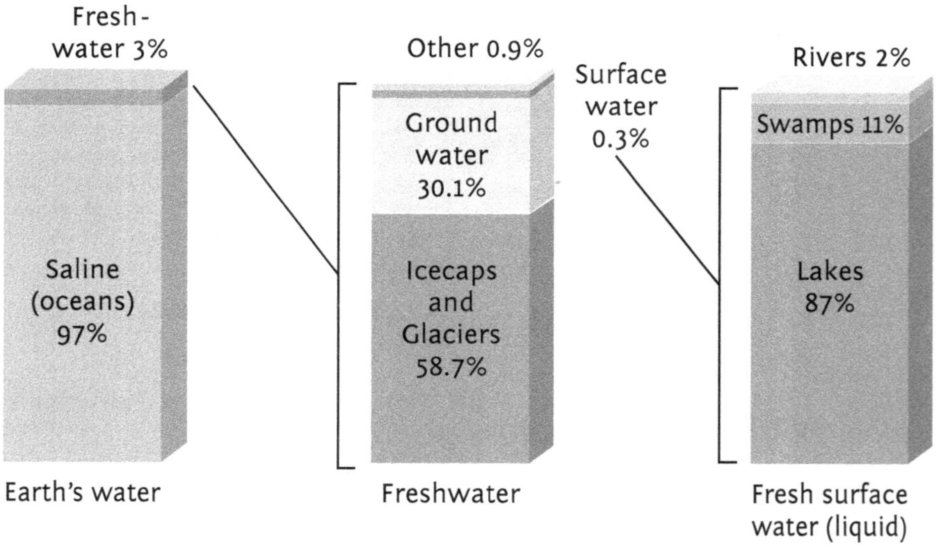

Source: US Geological Survey (1999).

Groundwater (as opposed to surface water such as lakes, swamps, and rivers) makes up approximately one-third of the Earth's freshwater supply. Freshwater, in turn, accounts for about 3 percent of the water found on Earth.

is being rapidly depleted in some areas by pumpage for municipal and agricultural uses (Alley, Reilly, and Franke 1999). At an average decline rate of two meters per year, this groundwater resource could be depleted in less than fifty years, with half the total volume of water gone by year 2020 (Upper Midwest Aerospace Consortium 2010).

Aquifer Characteristics

Aquifers comprise the invisible, subsurface part of the Earth's hydrologic (or water) cycle. Driven by radiation from the sun, the water cycle encompasses the combination of physical processes responsible for circulating water continuously through nature. Within this cycle, which has no definable beginning or end, water passes through three different phases: liquid, solid, and gas. In addition to the above-ground activities of precipitation, runoff, evaporation, and condensation, the water cycle includes groundwater recharge (infiltration), storage, and discharge.

There are two major types of aquifers. The first is where groundwater fills, or partially fills, a sediment or rock matrix and the upper surface of the saturated zone is free to rise and fall. This type is referred to as an *unconfined* (water table) aquifer. The second type is referred to as a *confined* (artesian) aquifer. (Figure 2.) Confined aquifers are capped by a layer of material that restricts groundwater from rising above the base of that layer, thereby creating pressure that can be released in a well that penetrates below the confining material. Such wells are referred to as artesian wells and may flow if the pressure is sufficient to cause the water to rise above land surface. A flowing artesian well can be important for providing water at land surface with little cost to adjacent neighborhoods. A relatively minor, third type of aquifer is the *perched* aquifer, which includes special cases of unconfined aquifers with particularly impervious bases of tightly cemented or otherwise impenetrable material; they occur mainly in arid environments.

Unconfined aquifers are generally present in the shallow subsurface and bounded at the bottom by impermeable material. Known also as water-table aquifers, because the top of the saturated zone in an unconfined aquifer is a water table (figure 2), unconfined aquifers receive recharge directly from the infiltration

Figure 2. Types of Aquifers and Age of Groundwater

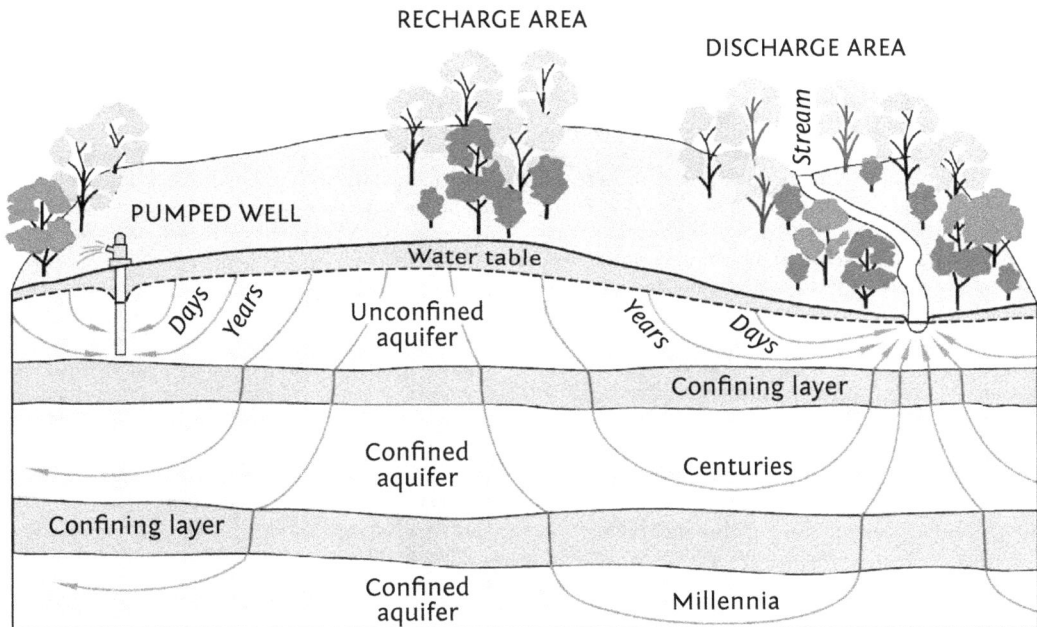

Source: Modified from US Geological Survey (1999) illustration.

There are two major types of aquifers: unconfined *(water table) aquifers and* confined *(artesian) aquifers. Because confined aquifers are comparatively deep and typically recharged in remote locations, it can take thousands of years for them to recharge naturally.*

of precipitation and surface water. The depth of a water table below land surface varies. In humid regions—particularly near streams, lakes, or oceans—the water table is generally within a few meters of land surface. Under different climatic or complex hydrogeologic conditions, however, groundwater depths can exceed hundreds of meters.

Confined aquifers are bounded, at the top and bottom, by relatively impermeable material called *confining layers*. Confined aquifers are characterized by groundwater under pressure that is equal to or greater than atmospheric pressure.

Groundwater Recharge, Discharge, and Storage

Aquifers are replenished primarily as precipitation falls upon the Earth's surface and percolates downward into the subsurface through a process known as aquifer recharge. In arid and semiarid regions, aquifers can be recharged by leakage from surface-water bodies. Recharge rates are controlled by a variety of factors, including soil type, nature of vegetation, antecedent soil moisture, and rainfall intensity.

In general, recharge to an unconfined aquifer originates from the land surface directly above the aquifer. In contrast, recharge to confined aquifers typically occurs tens or hundreds of kilometers away, generally at higher elevations where the aquifer rises to the surface. Once recharged, groundwater percolates down into confined parts of the aquifer. In addition to the importance of meeting water-quality standards, the degree to which groundwater remains a sustainable resource depends on the balance between rates of recharge and the rates of discharge, particularly those of pumping.

Unconfined aquifers are typically shallower than their deeper, confined counterparts; consequently, unconfined aquifers generally recharge more rapidly and are more vulnerable to pollution and water-quality degradation. Because confined aquifers are deeper and typically recharged in remote locations, confined aquifers can recharge so slowly that they are not necessarily renewable through natural recharge within a human lifetime.

From areas of recharge, groundwater generally flows along pathways of least resistance toward areas of discharge such as streams, lakes, and wetlands—all of which typically occupy topographic depressions. As a result, groundwater is a major contributor to streamflow, with a strong influence on lake levels and wetland habitats for plants and animals. Two common areas of groundwater discharge are springs and wells.

Springs commonly flow from rock outcrops on hillsides or along stream channels where migrating groundwater has encountered a relatively impermeable part of an aquifer and is diverted onto the land surface, or perhaps directly into a stream, lake, wetland, or ocean. In contrast to a surface spring, from which water either seeps (via gravity flow) or discharges under pressure, a water well is an excavation (of a usually narrow and vertical nature) into the subsurface, typically equipped with a pump that lifts groundwater from deep within an aquifer.

Groundwater that is neither recharging nor discharging from an aquifer is considered aquifer storage. Unless pumped out or withdrawn through other means, groundwater can remain in storage for days, years, or even centuries.

The age of groundwater in storage varies from aquifer to aquifer and from place to place, depending on its proximity to a recharging area. The age of groundwater in storage increases progressively along flow paths between areas of recharge and discharge. In shallow, local-scale flow systems, groundwater ages near areas of discharge can vary from less than a day to hundreds of years. In deep, regional-scale flow systems with long flow paths (tens of kilometers), groundwater ages may exceed thousands and approach tens of thousands of years.

Economics and Legislation

Different regions of the world approach regulation of use of groundwater by the development of special laws. Some of these laws may be derived from the Roman approach of "state ownership" and others are related to old English law of "rule of capture," where the landowner had an almost uncontrolled right to withdraw groundwater. Droughts and irrigation practices have led to changes in English law governing groundwater withdrawal. In general, the approach has paralleled that of some western US states, where riparian rights have been replaced with those of prior appropriation. Riparian rights refer to the rights of owners of land adjacent to water sources, in this case the groundwater under their land. Prior appropriation refers to historic use amounts being translated into government-apportioned rights. In the latter situation, no person may use water from a source without a license from some governmental entity, except some minor uses that make small demands on the water source, such as household or nonirrigation agricultural use. There are numerous other approaches to governing the use of water, including Spanish law, which was a system of grants from the government, an approach that has been used in lands once governed by Spain. The state of Texas and countries like Chile offer examples of the use of such laws. Countries such as Israel, where the climate is especially dry, may utilize a water code stating that water is public property, controlled by the state, intended for the use of its inhabitants and for the development of the country. In this type of code, permits are called "production licenses" and may only be used for specified purposes. Other types of law exist in Africa where, in many cases, tribal and clan uses are more important than individual use. Of all the laws widely used, western American prior appropriation has been cited as one of the best water use laws. The aforementioned Ogallala Aquifer, under the United States' High Plains, is still under "rule of capture" with some modification, such as well spacing rules. Western states, in contrast, do not follow the old English law of rule of capture. Prior appropriation allows for growth and adaptation, and its method of state control allows for planning and new uses. It's flexible enough to deal with the conditions of both humid and dry areas.

Control in the various laws varies from individual, tribal, and clan, to governmental entities and sometimes heads of government. The laws generally have provisions to deal with overuse, depletion, or special circumstances presented in different areas. Political influence on the type of law utilized is universal and the type of government may have a major role in the type of control.

The reduction or degradation of a surface-water supply, either actual or potential, as a consequence of groundwater exploitation can greatly complicate the equitable administration of water rights. Although there is no difference between the composition of surface water and groundwater, water laws in many countries do not necessarily recognize the physical connection between groundwater and surface water. An example of the sometimes convoluted nature of water management is the management of the Edwards Aquifer in Texas. Surface water in the contributing watershed is state water. It then recharges—becoming rule-of-capture water—and later resurfaces as spring water, becoming state water again. This demonstrates why water should be managed by an agency that can have jurisdiction over the entire watershed; in this case the state of Texas water agency may be the best managing authority. Groundwater development is restricted in many parts of the western United States with the objective of minimizing the impact on surface water; proving or compensating for such an infringement, however, is not a simple hydrologic or legal prospect.

Groundwater Exploitation

Unabated pumping from overexploited aquifers can damage local hydrogeologic and aquatic environments. Some of the more serious consequences of aquifer exploitation include groundwater depletion, well interference, surface-water infringement, water-quality degradation, and land subsidence.

Groundwater Depletion

Although the totality of groundwater depletion is not quantifiable throughout the world, evidence of overexploited aquifers is obvious in parts of Australia, India, Mexico, Thailand, North Africa, north China, the Middle East, South and Central Asia, and the western United States. According to the US Agency of International Development (USAID), groundwater is being depleted globally at the rate of at least 160 billion cubic meters per year, nearly twice the rate of Nile River discharge (USAID 2007). It is estimated that about seven hundred to eight hundred cubic kilometers of groundwater (five times the volume of the Dead Sea) was depleted from aquifers in the United States during the twentieth century (Konikow and Kendy 2005). USAID estimates that, during the next twenty-five years, India could lose 25 percent or more of its current crop production because of unsustainable rates of pumping (currently more than twice the rate of aquifer recharge), which are responsible for one to three meters per year of water-level decline (USAID 2007). Jordan and Yemen currently withdraw 30 percent more groundwater from aquifers every year than is being replenished (Hinrichsen, Robey, and Upadhyay 1997). According to USAID, Mexico is likewise depleting groundwater reserves in some agricultural regions at rates causing water-level declines of more than three meters per year (USAID 2007).

Well Interference

Well interference can occur from excessive pumping of improperly spaced water wells. Such interference is recognized when a well's water level and/or yield is significantly decreased as the direct result of pumping from another (typically nearby) well. (Figure 3.) Interference occurs as the individual cones of depression converge, thereby compounding the effect of pumping and increasing the three-dimensional distribution of drawdown. The degree of interference depends on the rates of pumping as well as the prevailing aquifer conditions and characteristics. Most of the western United States and other water-regulated regions enforce laws designed to minimize damage from well interference and the associated reductions of production in previously existing wells.

Surface-Water Infringement

Groundwater pumping can affect not only the availability and distribution of water for human consumption, but also the maintenance of instream flows important to animal and plant species and other environmental considerations. Pumping-induced variations in the time and/or spatial distribution of groundwater entering or leaving a stream can significantly affect the quantity and quality of surface

Figure 3. Cone of Depression

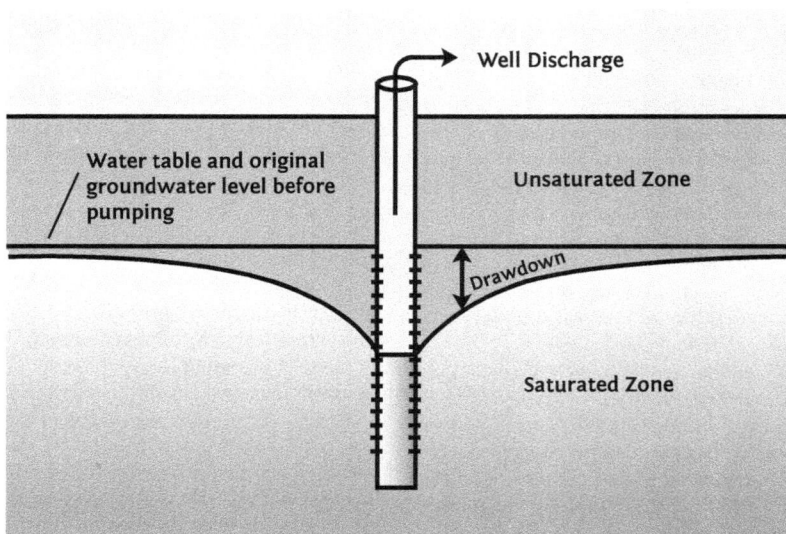

Source: Modified from a US Geological Survey (1999) illustration.

Cones of depression are the result of the pumping of a well drilled into the saturated zone of an unconfined aquifer. A well-managed aquifer will have wells that are properly spaced to avoid drawing water from each other, which increases the rate of depletion.

water. In addition to possibly affecting water acidity and temperature, changes in the nature of groundwater and surface-water interaction can produce shifts in nutrient and dissolved-oxygen concentrations that, in turn, may adversely affect the habitats of fish and other aquatic life.

Water-Quality Degradation

The use of almost any groundwater—whether for municipal, industrial, or agricultural purposes—typically results in increased temperature or increased concentrations of dissolved constituents, such as sodium and sulfur. The water quality of environments that receive contaminated return flow can be seriously impacted by groundwater exploitation. The problem with over-irrigated, water-logged soils can become particularly acute when salts leached from the top soil are concentrated through continued recycling of groundwater for irrigation. Such salinization can lessen agricultural production, degrade groundwater sources, and devastate irrigation projects. In coastal areas, where many of the world's largest cities are situated, excessive groundwater decline can allow seawater to creep inland and intrude freshwater zones, significantly reducing the volume of available potable water.

The removal of the most easily withdrawn fresh groundwater can leave behind water of inferior quality

due to leakage induced from land-surface bodies of contaminated or saline water. As groundwater is depleted, its place in the dewatered aquifer may refill with inferior water from hydraulically connected sources of toxins associated with factories, landfills, septic tanks, solid-waste dumps, mining operations, and leaky underground facilities or other receptacles of hazardous waste. From a public health standpoint, over-pumped aquifers are the most vulnerable to contamination from pesticides, nutrients, heavy metals, hydrocarbons, and toxic byproducts of the agricultural, manufacturing, processing, and transportation industries. Once polluted, groundwater sources are extremely difficult and expensive to restore.

Land Subsidence

The pumping of groundwater from a confined alluvial (sand and gravel) aquifer reduces the fluid pressure of the pore space between alluvial particles. Such decompression of aquifer porosity can induce a slow drainage of water from adjacent zones of confining material. If composed of compressible silt or clay, continued pumping can reduce water pressures in the confining material, causing it to compress from the weight of overlying strata. Where recharge is either unavailable or limited to the extent that the drained pore space remains unfilled, the compaction is typically translated through the subsurface to ground level—resulting in land subsidence, structural damage, and altered drainage patterns. Land subsidence that results from irreversible compaction of low-permeability confining material is permanent, which not only suppresses the land surface, but also greatly reduces the aquifer's capacity to store and transmit water.

Major subsidence resulting from excessive pumping has been recorded in several areas of the globe, including Bangkok, Thailand (1–2 meters); Houston-Galveston, Texas (1–2 meters); New Orleans, Louisiana (2 meters); Shanghai, China (2–3 meters); Venice, Italy (3 meters); Mexico City, Mexico (7 or more meters); and San Joaquin Valley, California (8 meters) (USGS n.d.).

Managing for Sustainability

Groundwater depletion is the inevitable and natural consequence of extracting water from an aquifer. Initially, pumpage is drawn from aquifer storage, but with time it is increasingly derived from decreased discharge and/or induced recharge. To minimize the extent of depletion and maintain potable groundwater reserves, some degree of sustainable management is prudent. According to the US Geological Survey, sustainable groundwater management means prioritizing the (1) effective use of groundwater in storage, (2) preservation of groundwater quality, (3) preservation of hydrologically connected aquatic environments, and (4) integration of groundwater and surface water into comprehensive (conjunctive-use) programs of water planning and management.

Innovations

Typically, the approach of most governments is to develop new water laws that provide the best solution to the sustainability of water supplies. For example, efficient use of water for irrigation by limiting the amount of evaporation is common; one technique that minimizes evaporative losses is drip irrigation. In some places, such as El Paso, Texas, and Riverside, California, wastewater discharges are highly treated so that they can be either reused for cooling or recharged to provide additional filtration in aquifers before being extracted again from wells. This approach may be useful in extending the life of an aquifer or keeping saltwater from intruding inland as freshwater is withdrawn.

Sustainable agriculture may also involve the development of new plant types that are able to use more saline water sources or survive on less water. Not all crops are equally efficient in supplying good nutritional resources, such as protein. The development of disease- or cold-resistant plants may also allow better distribution of agricultural production around the globe. The production of

plants to be used for biomass energy production is an area that is now receiving increased attention.

Several technologies, both natural and human made, aid in management of sustainability for a changing climate, including the following:

- *Watershed and Ecosystem Management.* Examples include low impact development (LID) and best management practices (BMPs). LID manages runoff at or near the source to enhance aquifer recharge. These controls should be in the form of BMPs designed to mimic pre-development hydrology; examples include retention basins, vegetated swales, and porous pavement and infiltration trenches.
- *Water Scarcity Issues.* Holistic review of water sources should include conservation, demand management such as nonpotable reuse (water that has been through a wastewater treatment plant and has received sufficient treatment to allow it to be used for nonpotable purposes; there are some areas that treat their reuse water to the point it can be polished through ponds and then treated for drinking), desalination, interbasin transfers, and conjunctive use of different supply sources.
- *Demand Management.* This increases the productivity of current water resources by focusing on more efficient use. Included here are initiatives such as flow control, leak prevention, metering, conservation rate structures, public education, and regulations.
- *Aquifer Storage and Recovery (ASR).* This involves recharging aquifers from other sources. The quality of this water must not degrade the aquifer. The water is stored when available and used when needed, such as during droughts. An advantage is that the stored water is not subject to evaporation.
- *Reclaimed Water.* This typically utilizes treated wastewater effluent and is used to replace potable water for landscape irrigation, cooling, and industrial uses.
- *Desalination.* This involves the removal of dissolved solids (salts) from seawater or brackish groundwater. The processes used are either thermal or membrane based. The former method uses different types of distillation and condensation of the water vapor. Thermal methods are typically used for highly saline water, such as seawater. Membrane processes may involve electrodialysis or reverse osmosis using electrical potential to drive ions through membranes with ion exchange resins. This technology is more often used for brackish water, especially if a low-cost source of electricity is available. Today the method may be coupled with solar generation of electricity (either panels or wind generated).

Boundary Considerations

Unlike the damming effect or containment aspects of a hydrogeologic boundary, the flow of groundwater does not begin or end at political boundaries; pumping on one side of a political boundary can dramatically affect the distribution and availability of groundwater on the other. Because groundwater moves according to the laws of physics rather than those imposed by humankind, sound aquifer management requires the cooperation and responsible participation of all affected governmental, political, and legal entities. To fully account for the resources of any given groundwater system, hydrogeologic investigations should strive to transcend local, state, and even national boundaries. Governmental agencies and managers of groundwater resources that extend across political boundaries should make every effort to coordinate their goals and strategies for the common interest of groundwater sustainability.

Controversies

In the United States, some of the most controversial actions involve situations where historic uses of water have diminished the availability of groundwater that provides habitat for endangered species. An example of this is the use of groundwater in the Nevada desert for borax production or mining activity, which causes drawdown in the Devil's Hole, a location of an endangered pupfish species. Another example occurs in the Edwards Balcones Fault Zone Aquifer in Texas, where the increasing use of this water source for agriculture and a large metropolitan area (San Antonio, Texas) has been associated with limiting the flow of large first order springs— springs that flow more than 2,800 liters per second—that are home to several endangered species. In this case, the approach to solving the problem involves the Edwards Aquifer Recovery Implementation Program (EARIP). This is a collaborative, consensus-based stakeholder process that seeks to balance groundwater development and use with the recovery of federally listed endangered species. Many stakeholders are working to develop a plan to protect the federally listed spring and aquifer species potentially affected by the management of the Edwards Aquifer. The goals of the plan include contributing to the recovery of these species.

Another controversy involves the contamination of high-quality water supplies as a result of inept or uncontrolled operation of oil and gas wells and/or the location of waste-disposal areas. An example of this issue comes from the oil industry, where in the past it was common practice to remove the casing from a well when it stopped producing sufficient quantities of oil. In numerous cases, this practice resulted in the contamination of high-quality aquifers with oil or brines commonly associated with oil production; in most cases the extent of the damage to the aquifer is such that it is not cost efficient to clean and repair.

It was not uncommon in the past, and even in some parts of the world today, to find landfills and other waste-disposal areas in places where the leachate (the water traveling through the waste and continuing down into aquifers) contained many contaminants such as nutrients, heavy metals, radioactivity, salts, and organic compounds that do not break down. Unfortunately, some of these may be carcinogenic. A common example of contamination is the case of nitrates, a group of compounds used as fertilizers or resulting from the breakdown of manure from confined animal operations, though other industrial sources of nitrates also occur. These can pose serious health risks for infants. Once the nitrates seep into groundwater utilized for drinking, they are consumed or used in the preparation of infant formula; they also occur in mother's milk as a result of the mother drinking the high-nitrate water.

Implications and Outlook

Aquifers play a vital role in supplying water for use by humans in many ways. In most areas, there is a need to modernize groundwater laws to meet current demand and prepare for the future sustainability of these supplies. Although it can be difficult, society must look to the future and consider how present actions may affect future resources. Pollution-control agencies should consider the protection of this valuable natural resource, especially as many actions that can contaminate groundwater are out of sight (being underground) and therefore "out of mind." Although groundwater comprises only about 1.7 percent of all Earth's water, it amounts to about 30.1 percent of all freshwater (Gleick 1996). Although climate change might ultimately impact the global distribution of most (if not all) water sources, the current consensus indicates little more than that some regions could receive more rainfall while others receive less (Lobell and Burke 2008). To circumvent the threats of depletion, pollution, and associated ramifications, objectives and methodologies must shift from developing new supplies to conserving, augmenting, and reallocating existing resources toward more beneficial uses (Molle 2003). New technologies are expected to help, especially those related to recovering polluted or saline sources through advanced methodologies for desalinization.

Although the causes of aquifer depletion and most associated implications are reasonably well understood, timely solutions are not always available. Hydrologists and water planners agree, however, that aquifer sustainability requires careful monitoring of the available resources and a responsible degree of cooperation among managers, developers, and users to establish and enforce pumping limits and water-quality standards. Rather than "first come, first served" approaches to groundwater development, nations must strive to protect the availability and quality of the resource to ensure that each competing interest has access to an appropriate share. As the world's population grows, so will the competition for water. As readily accessible water supplies continue to decrease, the demand for groundwater (particularly for drinking and agricultural purposes) will escalate because of the overuse and pollution of alternative resources.

Groundwater sustainability ultimately requires an appropriate combination of reduced consumption (conservation), increased recharge (augmentation), increased access to alternative water supplies (reallocation and recirculation), and/or reduced or eliminated sources of contamination. Although groundwater depletion and some degree of water-quality degradation are natural effects of groundwater production, an aquifer's viability is potentially sustainable through innovative methods of groundwater management, development, and use.

Education of the populace is important so that the public understands the interconnectivity of their actions. The American Water Works Association (AWWA) has developed the following definition: "Sustainability means providing an adequate and reliable water supply of desired quality—now and for future generations—in a manner that integrates economic growth, environmental protection and social development" (AWWA 2010). There is no more central resource for all than readily available clean water. Some of the cleanest water currently occurs in aquifers. If we overuse, contaminate, and fail to conjunctively manage this resource for the future, the cost to humankind will be much greater to obtain good, safe water.

Glenn LONGLEY and Rene Allen BARKER
Texas State University

See also in the *Berkshire Encyclopedia of Sustainability* Agriculture (*several articles*); Desalination; Water (Overview); Water Energy; Wetlands

FURTHER READING

Alley, William, M. (2003). Desalination of ground water: Earth science perspective. US Geological Survey, Fact Sheet 075–03.

Alley, William M.; Reilly, Thomas E.; & Franke, O. Lehn. (1999). Sustainability of groundwater resources. US Geological Survey, Circular 1186.

Alley, William M., & Leake, Stanley A. (2004). The journey from safe yield to sustainability. *Ground Water, 42*(1), 12–16.

AWWA Sustainability Initiatives Coordinating Committee. (2010). Committee report: AWWA takes a leadership role on sustainability. *Journal of the American Water Works Association, 102*(6), 81–84.

Bear, Jacob. (1979). *Hydraulics of groundwater.* New York: McGraw-Hill, Inc.

Bredehoeft, John D.; Papadopulos, S. S.; & Cooper, H. H., Jr. (1982). Groundwater: The water budget myth. In *Scientific basis of water-resource management, Studies in Geophysics* (pp. 51–57). Washington, DC: National Academy Press.

Driscoll, Fletcher G. (1986). *Groundwater and wells.* St. Paul, MN: Johnson Filtration Systems, Inc.

Freeze, R. Allen, & Cherry, John A. (1979). *Groundwater.* Englewood Cliffs, NJ: Prentice Hall, Inc.

Frias, Rafael E., & Binney, Peter D. (2010). Sustainable water resources technologies for a changing climate. *Water Resources Impact, 12*(4), 3–5.

Galloway, Devin.; Jones, David R.; & Ingebritsen, S. E. (2001). Land subsidence in the United States. US Geological Survey, Circular 1182.

Giordano, Mark. (2009). Global groundwater? Issues and solutions. *Annual Review of Environment and Resources,* 34, 153–178.

Gleick, P. H. (1996). Water resources. In Stephen H. Schneider (Ed.), *Encyclopedia of climate and weather: Vol. 2* (pp. 817–823). New York: Oxford University Press.

Heath, Ralph C. (1983). Basic ground-water hydrology. US Geological Survey, Water-Supply Paper 2220.

Hinrichsen, Don; Robey, Bryant; & Upadhyay, Ushma D. (1997, December). Solutions for a water-short world. Population Reports, Series M, No. 14. Baltimore: Johns Hopkins School of Public Health, Population Information Program.

Johnson, Corwin W., & Lewis, Susan H. (Eds.). (1970). *Contemporary developments in water law.* Austin, TX: Center for Research in Water Resources, University of Texas.

Kenny, Joan F., et al. (2009). Estimated use of water in the United States in 2005. US Geological Survey, Circular 1344.

Konikow, Leonard F., & Kendy, Eloise. (2005). Groundwater depletion: A global problem. *Hydrogeology Journal, 13*(1), 317–320.

Lobell, David B., & Burke, Marshall B. (2008). Why are agricultural impacts of climate change so uncertain? The importance of temperature relative to precipitation. *Environmental Research Letters, 3*(3), 034007.

Lohman, S. W. (1972). Ground-water hydraulics. US Geological Survey, Professional Paper 708.

Molle, François. (2003). Development trajectories of river basins: A conceptual framework. Research Report 72. Colombo, Sri Lanka: International Water Management Institute.

Remson, Irwin, & Randolph, J. R. (1962). Review of some elements of soil-moisture theory. US Geological Survey, Professional Paper 411–D.

ReVelle, Charles, & ReVelle, Penelope. (1974). *Sourcebook on the environment, the scientific perspective.* Boston: Houghton Mifflin Co.

Sophocleous, Marios A. (2000). From safe yield to sustainable development of water resources—the Kansas experience. *Journal of Hydrology, 235*(1–2), 27–43.

Theis, C. V. (1940). The source of water derived from wells: Essential factors controlling the response of an aquifer to development. *Civil Engineer, 10*(5), 277–280.

United States Agency of International Development (USAID). (2007). USAID Environment: Water—Groundwater Management. Retrieved January 11, 2011, from http://www.usaid.gov/our_work/environment/water/groundwater_mgmt.html

United States Geological Survey (USGS). (1999). The quality of our nation's waters—nutrients and pesticides. US Geological Survey, Circular 1225.

United States Geological Survey (USGS). (n.d.). Land subsidence: Bibliography of selected USGS references. Retrieved May 13, 2011, from http://water.usgs.gov/ogw/subsidence-biblio.html

Upper Midwest Aerospace Consortium. (2010). Fresh water: Groundwater overdraft. Retrieved January 11, 2011, from http://www.umac.org/ocp/GroundwaterOverdraft/info.html

Weeks, John B.; Gutentag, Edwin D.; Heimes, Frederick J.; & Luckey, Richard R. (1988). Summary of the High Plains Regional Aquifer-System Analysis in parts of Colorado, Kansas, Nebraska, New Mexico, Oklahoma, South Dakota, Texas, and Wyoming. US Geological Survey, Professional Paper 1400-A.

Winter, Thomas C.; Harvey, Judson W.; Franke, O. Lehn; & Alley, William M. (1998). Ground water and surface water: A single resource. US Geological Survey, Circular 1139.

Biocentrism

Biocentrism, or "life-centeredness," defines a perspective that played an important part in the development of ecological theory and ethics throughout the twentieth century. Because biocentrism focuses on the flourishing of life, its interdependence, and the need to protect ongoing evolutionary possibilities for all creatures, it provides a way of framing the goals of sustainability discourse and practice.

The term "biocentrism" means "life-centeredness." It indicates a central concern for and prioritization of the biotic community as a locus of value, and is most closely associated with the fields of environmental ethics and environmental philosophy. When paired with the concept of sustainability, biocentrism offers a means for deliberating about what is likely to promote the well-being of future generations, not just for humans but for all Earth-bound species.

In 1915 the horticulturalist Liberty Hyde Bailey (1858–1954) used the term "biocentric" in his book *The Holy Earth* to describe the "brotherhood relation" shared among all beings that "are parts in a living sensitive creation" (1919 [1915], 30). Bailey expressed his philosophy of "the oneness in nature and the unity of living things" throughout this work, but the title of the book alone indicated that he regarded the Earth and its life forms as sacred and worthy of reverent care. Biocentrism was thus early associated with responsibilities that went beyond material utilization of the Earth's resources.

During the twentieth century the science of ecology bolstered and provided content to the concept of biocentrism, highlighting the interdependence of natural systems and the importance of biotic diversity to their resilience. As the historian Roderick Nash put it, ecologists "were the scientists most likely to meet holistic-thinking theologians and philosophers half-way" (Nash 1989, 68). Beginning in the early 1970s, environmental philosophers who articulated what would become known as "deep ecology" drew attention to biocentrism as a critical facet of expanding ethical consideration to nonhuman species and the biosphere as a whole. The philosopher Arne Naess (1912–2009) is credited with naming "biocentric equality" (the right of all things to live and flourish) as one of deep ecology's most basic principles (Devall and Sessions 1985). Some have even promoted biocentrism as a religious alternative to human-focused "world" religions and secular humanism, arguing that biocentrism promotes "interspecies equality" and "a sense of planetary belonging" in which "the planet becomes its own church" (Watson 2005, 178).

Biocentrism is often contrasted with other terms that indicate the center of value (or the sacred center) for an individual or community. Anthropocentrism (human-centeredness) serves as the starkest contrast, and biocentrism is frequently paired with this term and used as a counterweight to suggest its ethical shortcomings. Terms like "ecocentrism" and "geocentrism" also highlight nonanthropocentric approaches to valuing the Earth and its living systems.

Critics of terms like "biocentrism" see them as overly abstract and unhelpful in offering unifying pragmatic solutions to environmental problems. But for some people this word continues to provide a shorthand way of articulating a larger philosophical vision in which all life is interconnected and which seeks to promote the interests of human and nonhuman life alike.

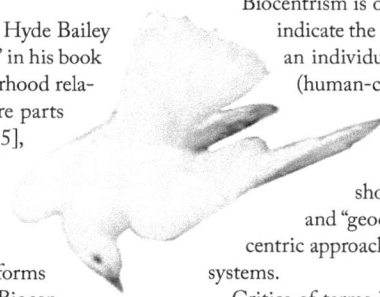

Biocentrism is one way of framing the goals of sustainability discourse and practice in that it focuses on the flourishing of life, its interdependence, and the need to protect ongoing evolutionary possibilities for all creatures. If sustainability is understood as a movement toward ensuring the local and planetary well-being of future generations, then biocentrism may be considered one method of rethinking the relationships between human communities and the natural world so that this vision might be approximated in practice.

Gavin VAN HORN
Southwestern University

FURTHER READING

Bailey, Liberty Hyde. (1919 [1915]). *The holy Earth*. New York: Scribner Press.

Devall, Bill, & Sessions, George. (1985). *Deep ecology: Living as if nature mattered*. Salt Lake City, UT: Gibbs Smith.

Lynn, William S. (1998). Contested moralities: Animals and moral value in the Dear/Symanski debate. *Ethics, Place, and Environment 1*(2), 223–242.

Nash, Roderick F. (1989). *The rights of nature: A history of environmental ethics*. Madison: The University of Wisconsin Press.

Norton, Bryan G. (1991). *Toward unity among environmentalists*. New York: Oxford University Press.

Rolston, Holmes, III. (1988). *Environmental ethics: Duties to and values in the natural world*. Philadelphia: Temple University Press.

Watson, Paul. (2005). Biocentric religion—a call for. In Bron Taylor (Ed.), *Encyclopedia of religion and nature* (pp. 176–179). London and New York: Continuum.

If moral strength comes from good and sufficient scenery, so does the preservation of it become a social duty. It is much more than a civic obligation. But the resources of the earth must be available to man for his use and this necessarily means a modification of the original scenery. Some pieces and kinds of scenery are above all economic use and should be kept wholly in the natural state. Much of it may yield to modification if he takes good care to preserve its essential features. Unfortunately, the engineer seems not often to be trained in the values of scenery and he is likely to despoil a landscape or at least to leave it raw and unfinished.

LIBERTY HYDE BAILEY

Source: Liberty Hyde Bailey. (1919 [1915]). *The Holy Earth*, pp. 116–117. New York: Scribner Press.

Buffers

A buffer is a protective barrier that can exist in natural or managed systems, such as urban areas or agriculture. In the latter, buffers often serve as a landscape design solution to human-made problems like soil erosion, surface water contamination, habitat fragmentation, and biodiversity loss; buffers have the potential to improve the health of human-dominated ecosystems. Obstacles to the use of buffers are primarily economic.

Buffers serve important functions in protecting our natural resources from the negative impacts of human activities. The term *buffer* refers to an entity that serves as a protective barrier, reducing or eliminating the flow of undesirable substances. In natural systems, vegetated buffer zones can protect rivers, wetlands, and other sensitive features from natural disturbances such as fires and floods. In the context of managed systems, buffers represent a landscape design solution that has the potential to reduce the impacts of a number of different anthropogenic problems such as soil erosion, surface water contamination, biodiversity loss, and habitat fragmentation.

Agriculture is one type of land use that can have detrimental impacts on our natural resources. Pesticides and fertilizers used in agricultural systems can be transported to rivers, lakes, and oceans, where they contaminate the water and harm aquatic organisms. Intensive soil disturbance from agricultural activities contributes to erosion and depletion of valuable topsoil. The expansion of agriculture across the landscape has resulted in a net loss of biodiversity, as natural habitats rich with a variety of species are replaced by homogeneous cropping systems. Urban areas, which are also managed systems, are not immune to these problems. In cities, lawns serve as sources of pollutants, building construction contributes

to soil erosion, and new development threatens diverse ecosystems. Buffers, when designed and located appropriately, have considerable potential to address these problems, offering a solution that could improve the health of human-dominated ecosystems.

Background

Historically, features such as riparian forests, wetlands, and hedgerows were retained and planted in agricultural landscapes, serving as buffers around cultivated fields and pastures. With the intensification of agriculture over the past 150 years, many of these features were removed and replaced with crops. Without noncrop habitats to divide fields and protect water resources, soil erosion from wind and rainfall became an increasing problem. The 1934 Dust Bowl, a great dust storm that occurred across the farmlands of North America, increased public awareness of the problem. In the years following that event, new soil conservation practices were introduced, including vegetated buffer zones within fields, along waterways, and beside roadways. Not until the 1970s, however, was the term *buffer* used to describe such features. At that time, conservation scientists began to study the effectiveness of buffers in improving water quality.

Today, buffers exist in a wide range of settings, offering a variety of functions depending on their configuration in the landscape and the composition of plant materials. *Riparian and wetland buffers* consist of perennial vegetation, oftentimes forest communities, located directly along water courses. These habitats can play a critical role in protecting water quality in agricultural and urban landscapes. *Wetlands* can themselves serve as very effective buffers, treating contaminated water before it enters rivers, lakes, or oceans. *Field margins* or *hedgerows* are located along the borders of crop fields in agricultural areas, and they serve an important function in reducing soil erosion and offering habitat for wildlife. *Windbreaks* or *shelterbelts* typically contain trees or shrubs to reduce wind erosion and protect crops, livestock, or homesteads from the harsh weather conditions. *Grass filter strips* are designed to intercept contaminants from storm-water runoff before it enters a water body. In urban settings, features such as vegetative swales, rain gardens, and constructed wetlands serve as buffers, if they are located between a source of storm-water runoff and a sensitive water body.

Benefits of Buffers

In the context of managed systems, buffers offer many benefits for the environment and society in general. They have the potential to reduce a number of negative impacts on our natural resources that result from current farming practices and urban development. Buffers reduce soil loss resulting from wind and water erosion, because the perennial plants with extensive root systems help to stabilize the soil and infiltrate water. Taller plantings, particularly trees, reduce the wind current that can carry uncovered topsoil. Buffers also protect water supplies by intercepting fertilizers, herbicides, heavy metals, and other contaminants from storm-water runoff from crop fields, residential lawns, or impervious surfaces. The mechanisms for treating storm water can include physical filtration of sediment-bound materials, chemical or biological transformation of materials in soils, and uptake of materials by vegetation. Performance in removing pollutants, therefore, varies substantially depending on the chemical structure and soil binding properties of the material. Research studies have shown buffers to be effective in capturing significant fractions of nitrogen, phosphorus, heavy metals, soil-bound herbicides, and organic materials.

While the primary function of buffers is typically to improve water quality by reducing erosion and intercepting pollutants, buffers offer many other environmental benefits. The perennial vegetation in buffers, often consisting of many different species, increases the biodiversity of flora and fauna, while also offering habitat for wildlife. With riparian vegetation, wildlife and aquatic organisms benefit from a more favorable microclimate (regulation of light and temperature), which provides greater access to food and water. Buffers can also serve as corridors to connect natural habitats and support the dispersal of organisms between fragmented patches.

In addition to the environmental benefits of buffers, they provide a range of other benefits for society. The water-quality benefits of buffers, for example, can improve the safety of drinking water and reduce the degradation of recreational waters. Buffers reduce flooding by infiltrating water and retaining flood water within wetlands. The result is a reduced hazard for people and less damage to built structures following flooding events. Trees and shrubs in buffers can also filter dust and unpleasant odors from the air, including around large livestock facilities. When they work as part of a greenway system, buffers can serve as corridors for wildlife as well as for people. They also offer important visual-quality benefits by greening the space, diversifying the landscape structure, and screening views of undesirable features. In addition to protecting recreational features, buffers can themselves provide recreational opportunities such as hunting, hiking, and bird-watching.

Barriers to Buffer Adoption

Despite the extensive benefits of buffers, many obstacles limit their widespread adoption. First is the cost of lost opportunities from competing land uses, many of which offer greater potential for profit. In rural areas, this is typically the yield from crops that would be grown on the area used for a buffer. Some farmers are tempted to use the buffer area for grazing livestock, but this can reduce the integrity of the buffer by compacting the soil, limiting the growth of vegetation, and allowing nutrient-rich manure to be deposited near sensitive sources (i.e., rivers). In urban areas, the values of competing land uses are even greater, with opportunities for residential or commercial development. This issue is particularly critical in the areas surrounding scenic lakes and rivers, where land values are often relatively high. Even in a floodplain, where development is not practical, nearby residents sometimes oppose the establishment of treed buffers because they obstruct the scenic view.

A second barrier to expanding the adoption of buffers is related to the direct costs of establishing and maintaining them. Some buffer types require expensive earth-moving equipment to grade the site to convey or retain water. For many buffers (particularly those located in riparian zones), native trees, shrubs, and herbaceous vegetation must be purchased to develop the appropriate plant community. Initial establishment may also involve labor and materials to install barriers to protect the young plants from grazing and browsing by wildlife. After the buffer is established, additional costs may be incurred by the maintenance of mowing, weed control, and sediment removal. The cost of establishing and maintaining a buffer may need to be covered by public funds, through subsidies to landowners or some other program.

This leads to the third barrier—the role of the government or nongovernmental agencies in allocating funds for buffers, which provide public benefits. In much of North America and Europe, programs have been established to subsidize buffers (along with other conservation practices), particularly in rural areas where agriculture is the dominant land use. Programs such as the Conservation Reserve Program (CRP) in the United States or agri-environmental programs in Europe are designed for this purpose. Agri-environmental programs reward farmers for environmentally friendly management practices such as establishing riparian zones or wetlands, enhancing hedgerows, and conserving areas with high biodiversity. Landowners, however, are often resistant to allowing the government to play a role in the land-use decisions, since the programs often involve multiyear contracts. In many other countries, nongovernmental agencies often play an important role in promoting buffers to protect natural resources by supplying landowners with planning tools, providing labor to help with establishment or maintenance, or even purchasing the land directly to convert to buffers. Even in that situation, landowners (particularly smallholders) may still be skeptical of strategies that appear to allow another organization to control the land-use decisions.

Landscape Design Considerations

The obstacles to expanding the use of buffers may be partially addressed through landscape design by focusing on opportunities to optimize buffer performance, while also considering the preferences of stakeholders. Important factors for buffer design include overall size, placement within the landscape, and selection of plant species. Water quality benefits, for example, will only be incurred if the system is designed to convey water through vegetation or retain water long enough to allow treatment. Riparian zones are often considered to be the best location for treating water before it enters a water course, but if the area of land has been constructed with tile drains, the water may bypass the buffer treatment system altogether. If the primary function of the buffer is to filter odor and dust from the air, the vegetation must be located downwind of the source area. In order to serve as corridors, the buffer should be designed to connect natural habitat areas. Overall, the buffer design will depend to a great extent on the primary functions to be fulfilled.

Not only are the ecological functions important to consider, but cultural and social functions might also be supported by good buffer design. These functions might include recreation, visual quality, education, artistic expression, and historic preservation. Recreational opportunities can be supported by integrating trails or pathways through buffers, and by establishing vegetation that would draw in birds and other wildlife for

bird-watching or hunting. Buffers designed to support the visual quality preferences of the landowners, nearby residents, and other stakeholders might be more widely adopted and protected for the long term. Where scenic views are important, for example, the buffers might be designed with openings consisting of short shrubs or herbaceous plants, allowing taller vegetation to frame the views. In agricultural landscapes, farmers and other landowners often prefer buffer designs that demonstrate an ethic of stewardship, with well-managed vegetation that reflects the organization of the cropping systems. Educational, artistic, and historic components might be integrated into the design, particularly in urban areas where it is important to have the support of many user groups to establish and maintain buffers. The cultural functions will depend to a great extent on the local landowners and residents, as well as the context of the site.

The regional context of buffers can play a very critical role in the success of the design, so consideration should be given to the primary environmental issues, competing land uses, and preferences of stakeholders. In tropical regions, for example, deforestation drives a very specific need to reduce erosion and protect water resources, particularly for vulnerable rural communities. There the focus is often on reestablishing stream corridor vegetation. In Africa, buffers have been used to reverse some problems created by land degradation from deforestation and improper use of herbicides, while at the same time trying to optimize systems so they do not compete with production of valuable food resources. In regions where much of the native vegetation was replaced with introduced species, such as parts of New Zealand, the interest in reestablishing native plants along streams and wetlands in urban and urban fringe areas has driven the design and establishment of buffers. Some regions have tied the establishment of buffers directly to locally important endangered species. Buffers have been promoted in the US Pacific Northwest to improve habitat and reduce pesticides that harm salmon, which are a highly valued species—both economically and ecologically. In Europe, hedgerows and other buffer types have been recognized and promoted for the positive impact they have on the aesthetics of the countryside landscape, which impacts agritourism in the region. Around the world, buffers have been used regularly to protect water resources, but their implementation is often more successful when tied to other specific goals that fit with the interests and preferences of the people living there.

Outlook for the Future

Several trends are likely to impact the adoption and design of buffers in managed systems in the future. First is the growing interest in developing landscapes to be sustainable and multifunctional. Buffers should be considered as a standard layer of rural and urban landscape planning. Even the design of individual buffer features can be conceived in a way that supports many more functions and synergizes conservation goals. Tree and shrub species in buffers, for example, might be selected to offer edible fruits and nuts that can be harvested by the landowner or users of the site. Increasingly, the interest in integrating biofuel production in the more sensitive and less productive areas has been considered. This strategy could be appropriate if the biofuel crops are native perennial plants with very low likelihood of invasiveness, and they would not be replacing another important plant community.

A second trend with buffers is to improve their performance (typically related to water quality) using advanced technologies such as specialized filters containing materials that have a high capacity for absorbing pollutants. Recycled steel slag, for example, has been used to remove high levels of phosphorus through subsurface-flow constructed wetlands. As new technologies for water treatment become available, they are likely to be integrated into the design of buffers. Finally, future strategies to increase the adoption of buffers will probably encourage greater participation from a wide range of stakeholders, including coordinated efforts among multiple landowners. Participatory planning approaches have been shown to encourage commitment from stakeholders, increase satisfaction with results, build trust in the process, and create more realistic outcomes. These trends, taken together, could have a significant, positive impact on the contribution buffers make in protecting our natural resources.

Sarah Taylor LOVELL
University of Illinois, Urbana–Champaign

See also in the *Berkshire Encyclopedia of Sustainability* Agroecology; Biodiversity; Biological Corridors; Brownfield Redevelopment; Community Ecology; Ecological Restoration; Habitat Fragmentation; Hydrology; Irrigation; Landscape Architecture; Landscape Planning, Large-Scale; Rain Gardens; Stormwater Management; Tree Planting; Viewshed Protection

FURTHER READING

Haycock, N. E.; Burt, T. P.; Goulding, K. W. T.; & Pinay, G. (Eds.). (1997). Buffer zones: Their processes and potential in water protection. *The Proceedings of the International Conference on Buffer Zones.* Harpenden, UK: Quest Environmental.

Hellmund, Paul C., & Smith, Daniel S. (2006). *Designing greenways—Sustainable landscapes for nature and people.* Washington, DC: Island Press.

Lovell, Sarah T., & Johnston, Douglas M. (2009). Designing landscapes for performance based on emerging principles in landscape ecology. *Ecology and Society, 14*(1), 44.

Lovell, Sarah T., & Sullivan, William C. (2006). Environmental benefits of conservation buffers in the United States: Evidence, promise, and open questions. *Agriculture, Ecosystems and Environment, 112*, 249–260.

Lowrence, Richard R.; Dabney, Seth M.; & Schultz, Richard. (2002). Improving water and soil quality with conservation buffers. *Journal of Soil and Water Conservation, 57*(2), 37A–43A.

Mayer, Paul M.; Todd, Albert H.; Okay, Judith A.; & Dwire, Kathleen A. (2010). Introduction to the featured collection on riparian ecosystems & buffers. *Journal of the American Water Resources Association, 26*(2), 207–210.

Meurk, Colin D., & Hall, Graeme M. J. (2006). Options for enhancing forest biodiversity across New Zealand's managed landscapes based on ecosystem modeling and spatial design. *New Zealand Journal of Ecology, 30*(1), 131–146.

Schultz, Richard C.; Kuehl, Amy; Colletti, Joe P.; Wray, Paul; Isenhart, Tom; & Miller, Laura. (1997). *Riparian buffer systems* (Publication Pm-1626a). Ames: Iowa State University Press.

Sullivan, William C.; Anderson, Olin M.; & Lovell, Sarah T. (2004). Agricultural buffers at the rural-urban fringe: An examination of approval by farmers, residents, and academics in the Midwestern United States. *Landscape and Urban Planning, 69*, 299–313.

United States Department of Agriculture—Natural Resources Conservation Service. (2000). Conservation buffers to reduce pesticide losses. Retrieved May 11, 2010, from http://www.in.nrcs. usda.gov/technical/agronomy/newconbuf.pdf

Yuan, Yongping; Bingner, Ronald L.; & Locke, Martin A. (2009). A review of effectiveness of vegetative buffers on sediment trapping in agricultural areas. *Ecohydrology, 2*(3), 321–336.

Clean Air Act

United States, 1970

Air pollution has plagued humankind since early history and escalated from a local to a global problem during the Industrial Revolution. In 1970, the United States addressed the problem nationally by implementing the Clean Air Act, designed to establish uniform standards for pollutants and their sources. Although the act has helped improve air quality, polluted air remains a global problem requiring further work to ameliorate.

The Clean Air Act (CAA) is the United States' primary legal response to the problem of air pollution. Such pollution long predated the enactment of the CAA in 1970; it was a problem in ancient times, then it became intolerable in London by the seventeenth century. The Industrial Revolution of the nineteenth century spread air pollution throughout Europe and the United States due to dramatic increases in the burning of coal and other fossil fuels. Then air pollution grew much worse during the twentieth century. Dense smoke was blamed for traffic accidents, blackened buildings, and dirty laundry in numerous cities early in the twentieth century. More ominously, dozens of people died and thousands became ill during severe episodes of air pollution in Glasgow in 1909; Belgium in 1930; England in 1931; Donora, Pennsylvania, in 1948; and London again in 1952. In the United States, the steel industry of Pittsburgh and the cars of Los Angeles earned both cities reputations for especially polluted air.

Air pollution was regarded as a local problem in the United States until the second half of the twentieth century. Numerous cities had passed laws addressing the "smoke problem" as early the 1880s, often with little success. (For instance, in Chicago, Cincinnati, and other cities in the Midwest that depended on high-sulfur bituminous coal, the penalties were negligible and hard to enforce.) Congress began to legislate on the subject in 1940, and

a series of early federal air pollution laws, such as the Air Pollution Control Act of 1955 and the Air Quality Act of 1967, sought to provide states with financial and technical assistance to solve the problem. But air pollution continued to worsen.

Congress passed the CAA in 1970, and amended it in 1977 and 1990. A massive law, the CAA was also historic. Unlike earlier laws passed by Congress, the CAA boasted uniform, national standards covering a wide range of pollutants and sources. Passed just one year after the United States put a man on the moon, the CAA reflected both the technological optimism of the times and the frustration with poor air quality in American cities. Much of the CAA can be understood as the product of what might be called "cooperative federalism," a dynamic balance between federal standard setting and state implementation.

The keystone of the CAA is its treatment of the most common pollutants of concern in the outside air, and most of the CAA's provisions are driven by its regulation of these "criteria pollutants." The Environmental Protection Agency (EPA) must establish national ambient air quality standards (NAAQS) for each criteria pollutant: carbon monoxide, lead, nitrogen oxide, ozone, particulates, and sulfur dioxide. The primary NAAQS must be set at a level to "protect the public health" with an "adequate margin of safety." Secondary NAAQS protect the public welfare, broadly defined to include effects on animals, wildlife, water, and visibility. NAAQS levels must be based solely on health considerations; the agency may not consider economic or technical feasibility. The EPA is required to review the NAAQS and make appropriate revisions at least every five years. In practice, however, the agency has been reticent to make any changes because of the tremendous political and economic costs involved.

Implementation at State Level

The NAAQS approach provides a classic example of inflexible but easily administered standards. Such "one-size-fits-all" standards are easier for an agency to establish, monitor, and enforce than the more flexible and tailored local standards that vary from place to place. In practice, the inflexible and uniform approach of the NAAQS has been tempered in implementation. The EPA sets ambient air quality standards nationwide, and each state then has the responsibility (and discretion) of setting emission standards that will result in attainment and maintenance of those standards. Each state is required to submit a state implementation plan (SIP) that demonstrates how the NAAQS will be achieved by the deadlines established in the statute. In principle, a SIP should satisfy the NAAQS while taking into account local conditions, thus allowing a degree of flexible, site-specific standards. In fact, the opportunity for local adaptation is even greater because there are 250 areas in which NAAQS are measured, known as air quality control regions (AQCR). In simple terms, a state creating a SIP must first inventory the current emissions from sources, choose control strategies for reductions, and then demonstrate through computer modeling that the SIP will satisfy the NAAQS levels.

Each state has only one SIP, but that SIP and its subsequent revisions often contain different provisions for each AQCR within the state. Generally, a state must first determine the existing level of air pollution and how that compares to the applicable NAAQS. Next a state decides how to limit or reduce pollution as necessary to achieve the NAAQS. A state needs to consider the many sources of pollution (such as industrial facilities, motor vehicles, and power plants), the types of regulatory responses (such as strict emissions standards, permit requirements, economic incentives,

and voluntary programs), and the appropriate means of enforcing the resulting rules, which might include offering tax incentives or mandating fines. Modeling plays a critical role in this process, with sophisticated computer programs attempting to predict how certain actions will affect future air quality. Each state crafts its SIP differently, and each SIP evolves to reflect changes in scientific understanding, technological advances, and public policy preferences.

Implementation at Federal Level

If the EPA believes the SIP will not achieve the NAAQS, the EPA may start a process that effectively supplants the SIP and requires the state to comply with the EPA's federal implementation plan (FIP). A FIP was developed for Los Angeles in the 1970s but was extremely controversial because it proposed curbs on land use and driving. Indeed, Congress took away the EPA's authority to impose a FIP on Southern California. Since then there have been occasional threats of other FIPs, but none have materialized.

According to the CAA, an AQCR must attain the NAAQS "as expeditiously as practicable, but no later than 5 years from the date such area was designated nonattainment." The EPA may extend that period for up to five years, and a state can request two additional one-year extensions. The 1990 amendments customized those deadlines for attaining the ozone NAAQS, giving marginal nonattainment areas three years to achieve the NAAQS and giving twenty years to extreme nonattainment areas. The CAA also provides for graduated attainment dates for areas that have not attained the particulate NAAQS, again depending upon the severity of the area's pollution.

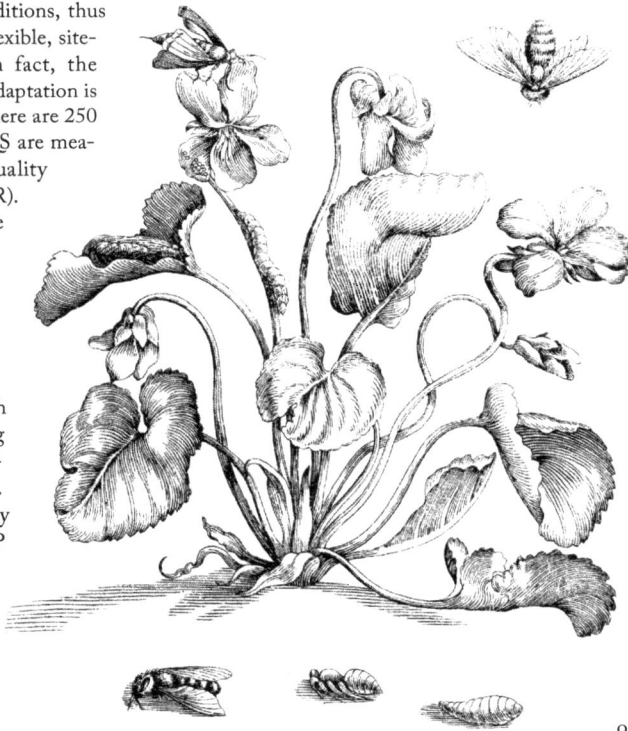

Amendments to the CAA

Congress neglected to say in the original 1970 version of the CAA what should happen if an area failed to attain the NAAQS established for a pollutant. Congress began to address the nonattainment problem in the 1977 CAA amendments, and then it added a number of new nonattainment provisions in the 1990 amendments. In particular, the 1990 amendments created a new approach to the nonattainment of ozone, carbon monoxide, and particulates. For example, there are five types of nonattainment for ozone—marginal, moderate, serious, severe, and extreme—with additional SIP requirements imposed as the nonattainment gets worse. The possible requirements include an inventory of emissions sources, increased emissions offsets, transportation and fuel restrictions, and the application of a "lowest achievable emissions reduction" for new sources of pollution. The areas that have been redesignated from nonattainment to attainment for ozone since 1990 must implement a maintenance plan for ten years in order to ensure that the ozone NAAQS will remain satisfied. The loss of federal highway funds is the most stringent sanction that the EPA may impose on states that fail to attain the NAAQS in a timely fashion. The EPA has been reluctant to sanction any states, especially through a politically charged cutoff of federal highway funds.

New Source Performance Standards

The SIP process allows states to impose restrictions on emissions from stationary sources such as incinerators, power plants, and industrial sites. Section 111 of the CAA, however, shields many of these sources from state control by requiring the EPA to set emission standards of harmful pollutants for new or modified stationary sources. These new source performance standards (NSPS) have been determined for over seventy categories of facilities and apply to "major sources" or major modifications. These emission limits are technology-based, reflecting the best pollution control technologies currently available in the industry. This indirectly ensures that costs will be taken into account, since presumably exorbitant control technologies wouldn't be commercially feasible. By its very name, the NSPS provision assumes that new and existing stationary sources should be treated differently. In this way, the CAA shifted the bulk of pollution control costs from existing businesses onto market entrants.

The CAA also discourages new polluters from locating in areas where the air is already clean. The prevention of significant deterioration (PSD) requirements require new sources in PSD areas to employ the "best available control technologies," which are at least as stringent, and usually more so, than the NSPS. The PSD program is complex.

In simple terms, it divides the country into three classes with varying degrees of restriction. Class I areas include national parks, such as the Grand Canyon; while Classes II and III areas cover the rest of the country. The class category determines the amount of development allowed through so-called growth increments. These increments place an upper limit on the increase in ambient concentration of pollutants in the area. The growth increment for particulates, for example, in Class I areas is 5 micrograms per cubic meter, while in Class II it is 19 micrograms per cubic meter. In other words, Class II areas can accept almost four times more increased emissions than Class I areas. When a new source or major modification seeks a permit to pollute, the source must calculate by modeling whether the increased pollution will violate the growth increment. Note that, as with nonattainment, the PSD requirements are pollutant-specific. An area can be a nonattainment area for ozone but a PSD area for particulates.

The CAA contains numerous provisions aimed at reducing pollution from cars and other motor vehicles. It requires the EPA to issue a series of rules to reduce pollution from vehicle exhaust, refueling emissions, and evaporating gasoline. It eliminates most uses of leaded gasoline. It requires cities with the worst ozone pollution to use gasoline that has been reformulated to reduce air pollution. It encourages the development of alternative fuels, such as natural gas, propane, methanol, ethanol, electricity, and biodiesel. It directs cities with air pollution problems to inspect individual vehicles to ensure that they are properly maintained. And the CAA provides that transportation projects, such as construction of highways and transit rail lines, cannot be federally funded or approved unless they are consistent with state air quality goals.

Hazardous Air Pollutants

The CAA contains a separate regulatory scheme for hazardous air pollutants. These are pollutants that could be extremely harmful even when released into the air in very small doses. The disaster that killed four thousand people in 1984—when Union Carbide Corporation, a multinational pesticide manufacturer operating in Bhopal, India, released methyl isocyanate into the air—demonstrates the immediate and deadly effects of exposure to airborne toxic materials. The 189 hazardous air pollutants include such substances as asbestos, benzene, hydrochloric acid, and vinyl chloride. But the history of the CAA's regulation of hazardous air pollutants has not been nearly as successful as the six criteria pollutants. The EPA struggled to identify the relevant pollutants and how to control them until Congress took matters into its own hands in 1990, expressly listing the 189 hazardous air pollutants and directing the

EPA to promulgate the maximum achievable control technology for regulating them.

In 1990, Congress added the Acid Rain Emissions Trading Program to the CAA. Acid rain is precipitation that contains sulfuric or nitric acid. Typically, acid rain results from the emission of sulfur dioxide and nitrogen dioxide from electrical power plants that burn lots of coal, especially eastern and midwestern coal with a high sulfur content. Once those pollutants are emitted, they can be carried for hundreds or even thousands of miles until they mix with moisture and fall back to the Earth as acid rain. In the United States, that often means that power plants in the Midwest are responsible for acid rain along the East Coast. Acid rain changes the chemical composition of lakes and rivers, often to the detriment of the wildlife, and it can harm forests as well. The injuries to humans that are attributable to acid rain include respiratory problems and property damage.

The Acid Rain Emissions Trading Program sets a permanent cap on the total amount of sulfur dioxide (SO_2) that may be emitted by electric power plants nationwide. Plants may only release the amount of sulfur dioxide equal to the allowances they have been issued. Each allowance is worth one ton of sulfur dioxide emissions released from the plant's smokestack. If a plant expects to release more sulfur dioxide than it has allowances for, it must purchase more allowances or use technology and other methods to control emissions. A plant can buy allowances from another power plant that has more allowances than it needs to cover its emissions. As of 2005, emission reductions were more than 7 million tons from power plants, or 41 percent below 1980 levels.

Greenhouse Gases

Much air pollution is local. Pollutants emitted in one place usually cause more harm there than in areas that are farther away. There are some exceptions to that scenario, though, as seen in the CAA's regulation of the sources of acid rain. But the most significant instance of global air pollution—and the point of greatest current controversy regarding the CAA—involves the emission of greenhouse gases that contribute to climate change. There are several differences between greenhouse gases, such as carbon dioxide, and traditional environmental pollutants. Carbon dioxide (CO_2) is naturally occurring, necessary for life, and even instrumental in the development of human welfare. It produces different effects than most air pollution, and it produces them indirectly. Many air pollutants cause respiratory ailments when they are inhaled, or they irritate people's eyes, or they interfere with the aesthetic enjoyment of desired views. In each instance, harm is caused by direct exposure to the pollutant. But carbon dioxide and other greenhouse gases trap heat from escaping the atmosphere, which results in the heating of the Earth's surface and other changes to the climate that injure the environment and harm human communities and wildlife alike.

In 2007, the Supreme Court ruled in *Massachusetts v. Environmental Protection Agency* that greenhouse gases are pollutants within the meaning of the CAA. Justice John Paul Stevens emphasized that the CAA's "sweeping definition of 'air pollutant' . . . embraces all airborne compounds of whatever stripe." Carbon dioxide and other greenhouse gases "are without a doubt 'physical [and] chemical . . . substance[s] which [are] emitted into . . . the ambient air.' The statute is unambiguous." Since the Court's decision, the regulatory battle has shifted to the EPA and to Congress. Numerous congressional hearings have explored the ramifications of the Court's decision, including one in which House Energy and Commerce Committee Chair John Dingell faulted the Court for creating "a glorious mess" that was not intended by the Congress that enacted the CAA. Meanwhile, the EPA has received more petitions asking it to regulate other activities under the CAA, including nonroad engines, shipping vessels, and aircraft. Soon after the Obama administration took office, the EPA made the requisite finding that greenhouse gases endanger public health and welfare, and the agency proposed to the CAA to regulate carbon dioxide. That proposal, in turn, prompted congressional efforts to prohibit the EPA from treating carbon dioxide as a pollutant under the CAA.

In each of these venues, the proponents of CAA regulation cite the law's effectiveness in reducing other types of air pollution. They also insist that the law contains abundant flexibility to allow the EPA to mold it to the circumstances presented by greenhouse gas emissions. Most potentially regulated parties see a much grimmer regulatory future if the CAA is employed to control carbon dioxide emissions. They refer to the "alarming consequences" of using such a flawed, unsuitable, and potentially destructive instrument as the CAA. They fear that the EPA will micromanage the entire US economy, force businesses to move overseas, compromise energy independence, and achieve relatively little environmental progress. The two sides also dispute whether the EPA has the power to avoid those consequences, given several recent US Court of Appeals District of Columbia Circuit decisions in which the court held that the plain language of the CAA precluded the EPA's efforts to develop flexible air pollution control programs.

Air pollution has been reduced dramatically in the United States since Congress enacted the CAA in 1970. The EPA's studies show that levels of each of the six criteria pollutants are declining, as are acid rain and many toxic air pollutants. Even so, the American Lung Association's *State of the Air 2010* report found that "the air quality in many places has improved, but that over 175 million people—roughly 58 percent—still suffer pollution levels that are too often dangerous to breathe." The CAA may be amended yet again to keep pace with the long-term effects of air pollution.

John Copeland NAGLE
Notre Dame Law School

See also in the *Berkshire Encyclopedia of Sustainability* Climate Change Disclosure—Legal Framework; Climate Change Mitigation; Convention on Persistent Organic Pollutants; Environmental Law—United States and Canada; Kyoto Protocol; *Massachusetts v. Environmental Protection Agency*; Montreal Protocol on Substances That Deplete the Ozone Layer; National Environmental Policy Act; Polluter Pays Principle; Trail Smelter Arbitration (*United States v. Canada*)

FURTHER READINGS

American Lung Association. (2010). *State of the air 2010*. Retrieved November 26, 2010, from http://www.stateoftheair.org/2010/assets/SOTA2010.pdf

Environmental Protection Agency (EPA). (2008). The plain English guide to the Clean Air Act. Retrieved November 18, 2010, from http://www.epa.gov/air/caa/peg/

Environmental Protection Agency (EPA). (2010). Our nation's air—status and trends through 2008. Retrieved November 18, 2010, from http://www.epa.gov/airtrends/2010/index.html

Heinzerling, Lisa. (2007). Climate change and the Clean Air Act. *University of San Francisco Law Review, 42*(1), 111–154.

Krier, James E. (1971). The pollution problem and legal institutions: A conceptual overview. *UCLA Law Review, 18*, 429–430.

Lazarus, Richard J. (2004). *The making of environmental law*. Chicago: University of Chicago Press.

Martineau, Robert J., Jr., & Novello, David P. (Eds.). (2004). *The Clean Air Act handbook* (2nd ed.). Chicago: American Bar Association.

Nagle, John Copeland. (2010). Climate exceptionalism. *Environmental Law Review, 40*(1), 53.

Oren, Craig N. (1988). Prevention of significant deterioration: Control-compelling versus site-shifting. *Iowa Law Review, 74*(1), 1–114.

Reitze, Arnold W., Jr., & Lowell, Randy. (2001). Control of hazardous air pollution. *Boston College Environmental Affairs Law Review, 28*(2–3), 229–350.

Rodgers, William H., Jr. (1994). *Environmental law* (2nd ed.). St. Paul, MN: West Publishing Company.

Squillace, Mark S., & Wooley, David R. (1999). *Air pollution* (3rd ed.). Cincinnati, OH: Anderson Publishing Company.

Strengths and weaknesses of regulating greenhouse gas emissions using existing Clean Air Act authorities: Hearing before the Subcommittee on Energy and Air Quality. US Congress. House Committee on Energy and Commerce. 110th Cong., 2nd Sess., 418 (2008).

Massachusetts v. Environmental Protection Agency, 549 US 497 (Mass. Ap. Ct. 2007).

Coastal Management

The conceptual basis for coastal management has evolved from sector-based approaches to conflict reduction to proactive consideration of trade-offs through spatial management and ultimately to coastal ecosystem governance, which advances human objectives with cognizance of limits imposed by natural systems. Enhancing sustainability may occur through reinterpretation of actions under the sector-based management system, through incremental changes to it or, alternatively, through fundamental changes that emphasize ecosystems and the services they provide to advance human welfare.

Coastal ecosystems include land and sea environments as well as the people who inhabit and use them. One way to define the geographic extent of this region is by elevation. Some describe the coastal zone (simply defined as the place where the land meets the sea) as the area ranging from the 200-meter land-elevation contour to its 200-meter-depth equivalent (Crossland et al. 2005). In many areas this zone would extend inland 70–100 kilometers and offshore to or just beyond the shelf break on the continental margin. Furthermore, for research purposes the extent of the zone would be expanded to encompass additional area to consider material flows (water, sediments, nutrients, contaminants) and social systems, which do not neatly fit within the elevation contours. In certain settings, coastal airsheds can be important too. Materials traveling by air settle to land and water areas of the coast, and attention to their sources becomes a part of research or management activity.

This land-sea geographic region has important biological properties (Day et al. 1989). Estuarine habitats show primary production levels that are among the highest on the planet. Estuarine food webs rest on phytoplankton, detritus, and submerged aquatic vegetation and produce very high abundances of benthic organisms per unit area. The fish community present varies in response to cycles of spawning, migration, and feeding in addition to year-to-year changes in absolute abundance. The diet of a single species of fish may include twenty food types, and as a result fish have important impacts on lower trophic levels through predatory pressure. These linked ecological systems are both sensitive to anthropogenic changes and important to human well-being through food and a variety of other services.

The land portion of the coastal zone is a highly desirable place to live and work. In fact, as of 2004, 53 percent of the US population lived in coastal counties, which account for only 17 percent of the land area of the country (Crossett et al. 2004). This means that coastal counties contain three hundred people per square mile, whereas the national average is ninety-eight individuals per square mile. In the New York coastal area, densities run as high as almost thirty-nine thousand people per square mile (Crossett et al. 2004). The coastal zone also includes the social system of its inhabitants, who bring values and expectations as well as an elaborate and growing system of laws, practices, and behaviors that structure how people relate to the natural systems they inhabit.

The political boundaries that are superimposed on this land-sea ecosystem vary by country, and the US system exemplifies the types of jurisdictions that can be established. Coastal counties are parts of states. Towns and municipalities reside within counties, and this nested system is subject to a variety of federal authorities. These jurisdictional boundaries are for the most part completely independent of the natural boundaries, such as watersheds. Nature thus is working with one set of boundaries while political jurisdictions operate with another ratio-nale for area delimitation. Coastal management must address this incongruity if it is to be successful. In the United States, the federal Coastal Zone Management Act encourages states to designate a management zone that extends outward from the land–sea boundary 3 miles (or in some instances 3 leagues) offshore. Outside of state waters the federal government manages a 200-mile exclusive economic zone.

This combination of environmental processes—a high and growing density of people, increasing anthropogenic alterations of the coast, and multiple governmental jurisdictions—creates difficulties for the managers of these complex coastal systems and for the larger societies. Coastal management as it is practiced today consists of three distinct approaches (Burroughs 2011).

Sector-based management is managing in response to separate activities (e.g., dredging, waste disposal, sprawl, oil development) that have an impact on others (e.g., rec-reation, fishing, tourism, aesthetics). This approach attempts to retroactively address the effects of one use on another use.

A second approach, known as *spatial management,* is to plan by geographic area either on land or at sea. In this process, the planning entity identifies uses that will be compatible in the same area or in areas adjacent to one another. This method looks across uses and seeks to avoid conflict largely by planning the location of activities. By virtue of its proactive approach, spatial management can reduce conflict, delay, and uncertainty as coastal lands and waters are developed.

A third approach, *coastal ecosystem governance*, adopts proactive planning within the context of both human needs and ecosystem capacity. Coastal ecosystem gover-nance promotes sustainable natural and social systems through diverse management techniques that shape human activity in concert with the limits of the natural system. This third approach provides the strongest foun-dation for sustaining coastal systems.

Sustaining Coasts through Ecosystem Management

Coastal societies face a continuing conundrum concern-ing conflicting goals for the land/sea region. The conflict is aptly captured in the 1972 US legislation for coastal management that states both the protection and devel-opment of the coast as goals. Fifteen years later, the Brundtland Commission raised additional challenges for coastal managers when it defined sustainable devel-opment as the ability of humanity "to ensure that it meets the needs of the present without compromising the ability of future generations to meet their own needs" (WCED 1987). The commission thereby established sustainability over long time periods as a measure on which to determine action. For example, actions in a coastal region are unsustainable if contaminated runoff is allowed to render coastal waters uninhabitable for valued organisms.

The dimensions of coastal management include what is to be sustained (nature, life support, and human com-munity), and what is to be developed (people, economy, and society) as noted by the US National Research Council (NRC 1999). Furthermore, a time horizon and the linkage between sustaining and developing are implicit in whatever solutions are adopted. In this frame-work, the difficult trade-off for the coast is between nature and economy. The NRC wisely asks whether the link should be nature only, nature mostly, nature but, nature and, or nature or economy; articulating these alternatives highlights the difficult choices to be made. In this context the central challenge of coastal management is determining the values associated with common inter-ests as well as creating and implementing coastal pro-grams that implement the type of sustainability most support. Historical records of coastal environments show this to be difficult.

Lotze et al. (2006) demonstrate that anthropogenic transformation of coastal systems can be tracked over thousands of years by examining species richness and associated indicators. Almost without exception, in the twelve predominantly Atlantic systems examined, the researchers found that important species and habitats (sea grass, wetland) have declined substantially. Nature in these settings has not been sustained as human societies moved from hunter-gatherer, to agricultural, and on to market-colonial periods. Other researchers (Fischer et al. 2007) envision the task of management as returning the biophysical, social, and economic system to a state that is sustainable, which means that human societies and econ-omies fit within the capacity of the Earth to support life through the foreseeable future.

Since governance systems for coastal regions seldom invoke sustainability directly and never provide specific instructions concerning which elements (nature, life sup-port, human community, people, economy, or society) are to be given precedence, creating and implementing man-agement systems remains difficult in the absence of explicit goals and authority.

Competing objectives for collective coastal action may be viewed in terms of ends and means (Burroughs 2011). As pressures on resources increase and societal values shift, the end or objective changes. Those values are quite different from the pre-1970s practices of filling or draining marshes and dumping wastes in coastal waters in primary pursuit of development goals. In fact the value shifts ultimately resulted in current discussions of ecosystem governance as a means to satisfy new goals.

Managing for long-term sustainability ultimately relies on the ability to create, adopt, and provide effective ways of governing human behavior. Orchestrating these policy elements to achieve sustainable coastal systems, both human and natural, results in positive sustainability trajectories (Burroughs 2012).

Coastal Ecosystem Governance: The Basis

For many, ecosystem management has become a way to reconcile the value conflicts discussed above. Ecosystem principles—properly understood—support sustainability. Ecosystems include people. This move toward ecosystem-based management as a means of refining both the definition and implementation of sustainability for coastal environments is evident in the results of a recent national commission in the United States. The US Commission on Ocean Policy (2004) represented many interests and provided a national perspective. The commission declared the need to manage all components of the ecosystem—including people—collectively. This principle has widespread support but is only sporadically reflected in current governance or policy directives.

Nonetheless, the potential for ecosystem-based management is great, and many have detailed what it would entail (Burroughs 2011). Statements about ecosystem-based management establish a goal of healthy, productive, resilient ecosystems able to provide services people need and want. Institutions with the capacity to manage holistically using the principles of ecosystem health, sustainability, and precaution are required. Such institutions, if present, would be particularly helpful in resolving management of

issues related to dead zones because administrators would be empowered to manage agricultural practices with full cognizance of the impacts on marine systems.

As summarized by Richard Burroughs (2011), individual authors emphasize the value of a common goal under which multiple sectors of coastal activity may be managed in an integrated fashion. Futhermore, natural boundaries with management regimes fitted to them (as opposed to political jurisdictions) are important for success. Collaborative planning built on public participation and equity should lead to flexible and adaptive program implementation. Ideally ecosystem-based management will account for cumulative impacts, incorporate precaution, and promote appropriate trade-offs among services. As more experience is gained with these concepts, they will inevitably be refined.

How much change is really required to implement ecosystem-based coastal management successfully? Burroughs (2011) considers three approaches. One approach would be to maintain the *conventional management* system, characterized by sectoral management, and simply try to improve its techniques to meet the new ecosystem goals. Unfortunately, the record suggests that this strategy would be difficult. In spite of many enhancements over the years, important measures of environmental health such as water quality are at best stable or declining. In fact, the increasing size of the low oxygen zone in the Gulf of Mexico makes clear that orchestration of fragmented government initiatives to control nutrient flow have not been adequate. There is also the matter of intent. Some purported transformations are mere rhetoric, rather than a substantive change in policy. By simply declaring current practices a form of ecosystem management, officials can avoid making needed changes.

A second approach is to undertake *incremental* change. This strategy involves changing parts of the existing policy system in small ways and then evaluating the consequences before contemplating additional changes (Lindblom

1959). The idea is that by "muddling through" officials can keep personal risk low and still transform the system—at least in part. The technique is not fully rational because managers do not specify all values and policy alternatives before selecting one. Rather, analysts agree on plausible actions but do not explicitly note the policy objective or indicate that they have selected the best means to reach it. In this ambiguous situation some elements of ecosystem-based management may find a home. The policy solution is selected by considering experience with small policy changes that have been made in the past. An important test for an incremental solution is whether it has moved forward without overwhelming objections. By muddling through, only part of the goal can be achieved, which means that the process must be repeated endlessly. In an incremental change, officials could utilize ecosystem thinking within the confines of existing legal structures and routines to enhance results. Almost all examples of current ecosystem-based management fall under incremental change.

In fact, incremental change has reshaped coastal and ocean management in the past. The arrangements for offshore oil development, originally established in the 1950s, were altered in the 1970s (Juda 1993). Public attitudes about the values of marine waters had shifted, and the US legal regime reflected this change. The National Environmental Policy Act, the Coastal Zone Management Act, and the revised Outer Continental Shelf Lands Act instructed government officials to consider the environmental consequences of decisions. The growing importance of environmental impacts was confirmed through judicial decisions and, together with changes in the law, altered the collection and use of environmental data (Burroughs 1981). After the transformation the government was increasingly charged with anticipating the detrimental environmental impacts of offshore oil development and taking action to avoid them by suspending part or all of certain developments.

A third possibility is that change becomes revolutionary—that is, both rapid and fundamental. *Fundamental change* becomes particularly important when the characteristics of the problem demand bold solutions, or when a shift in goals requires action that is beyond the reach of incremental change (Birkland 2005; Cortner and Moote 1999). Responding to the Great Depression, landing a human on the moon, and protecting equal rights required fundamental changes in policy. Coasts and oceans are receiving similar transformative attention by many today. The chair of the Pew Oceans Commission stated that their group found the ocean in crisis and advocated for "a fundamental change in this nation's posture toward its oceans," noting that "reforms are essential" (Pew Oceans Commission 2003, i). Just as the early 1970s were a time of rapid and fundamental changes in environmental policy (see the discussion of the Clean Water Act in the section below), ecosystem-based management could inspire similar advances in the twenty-first century.

Coastal Ecosystem Governance: The Practice

What would a fundamental shift to ecosystem-based management look like? Issues to consider include the rationale for such a change, the policy elements involved, and an example of a proposed change of this magnitude.

Rapid and dramatic policy transformation is not unprecedented. For example, the Federal Water Pollution Control Act Amendments of 1972, commonly known as the Clean Water Act, represented a profound shift from reliance on voluntary pollution-control measures to a "command and control" approach. To protect recreation, fisheries, and the environment, the new law set up a permitting system to control discharge of effluent from pipes, with treatment requirements based on technological feasibility. The act specified discharge limits for individual contaminants and created a monitoring and enforcement system to ensure compliance.

When pursuing change of this magnitude, it is important to consider both the principles driving it and the means of implementation. First as Robert Costanza et al. (1998) have noted, stakeholders should be involved in formulating and implementing policies, and those policies should be ecologically sustainable and socially equitable. Second, institutional scales for decision making should match the ecological setting. Third, potentially damaging activities should be approached with caution, and there should be ample opportunity to adapt and improve policies. Finally, Costanza and his colleagues recognized that sustainable governance of oceans rests on full allocation of social and ecological costs and benefits. If markets are used as a means of determining best policies, they will have to be adjusted to reflect full costs.

Changing to ecosystem-based management to enhance sustainability requires a shift in how we conceive the process of reaching decisions. In constitutive decision making, basic allocations of authority and control are made (Lasswell 1971; Clark 2002). The ideal constitutive decision process for ecosystem-based management will give precedence to the common interests that support human and natural systems while producing adequate authority to ensure implementation. The reliance on stakeholders in the planning process and proposals for adaptive management exemplify constitutive changes.

The central challenge of ecosystem-based management is to consider all factors that affect ecological systems and to manage activities so as to sustain the services provided (Levin and Lubchenco 2008). Ecosystem services are defined as the "conditions and processes through which natural ecosystems sustain and fulfill human life" (Daily 1997, 3). The Millennium Ecosystem Assessment divides ecosystem services into four categories (UNEP 2010, 8). The products or goods obtained from ecosystems are known as *provisioning services*. In coastal environments they include fisheries, mariculture, genetic diversity, medicines, transportation, minerals, and energy from wind, tides, and waves. Ecosystems also provide *regulating services* by controlling climate through carbon storage and land cover, water quality through decomposition, and natural hazards through protective features such as marshes and reefs. In addition, ecosystems also provide spiritual enrichment and aesthetic and recreational values, which are known as *cultural services*. Finally, the indirect and long-term benefits that are the basis for the production of all other ecosystem services are known as *supporting services*. Water or nutrient cycling and photosynthesis are examples.

To what extent and how might explicit consideration of ecosystem services advance coastal management? A change of that type would reorient the discussion of governance techniques to rely on principles of ecosystem-based management. Several investigators have laid out a conceptual basis for it (Slocombe 1998; Yaffee 1999; Layzer 2008; McLeod & Leslie 2009). The Millennium Ecosystem Assessment (UNEP 2006), among others, proposed that ecosystem services are an effective means to convert concepts into practices. Because governmental programs are not designed to protect and enhance ecosystem services in a direct manner, implementation of this approach will almost certainly require use of new tools. Pertinent tools include markets, tradable pollution permits, government ownership, government regulations, incentive payments, voluntary payments, and other means (Brauman et al. 2007; Ruhl, Kraft, and Lant 2007). Ecosystem services require thought and action based on natural systems, not political jurisdictions. Working at the regional scale will almost certainly lead to better understanding of cumulative impacts and to actions less likely to jeopardize the ecosystem services at stake.

Ecosystem service districts are one way to achieve this new goal (Heal et al. 2001; Lant, Ruhl, and Kraft 2008). The United States has many regional organizations that manage geographic areas for specific purposes. State law or local initiative can create a district to manage human behavior associated with watershed health and services, such as erosion, water supply, or floods. These serve as prototypes for ecosystem service districts, which can be linked in ways that allow their geographic jurisdictions to match ecosystem properties, thus ensuring effective oversight. An ecosystem service district is a governmental entity managing a geographic region and empowered to coordinate, zone, and tax. Hypothetically the ecosystem service district would select the least costly means of providing a service. For example, New York City faced a tradeoff between managing its watershed to protect drinking-water quality and installing water treatment. By establishing a prototypical ecosystem service district surrounding the reservoirs, the city could and did benefit from the ability of ecosystem processes to purify water in lieu of technology investments to achieve the same results (Heal et al. 2001). Sorting out the advantages and disadvantages of different proposals becomes more complicated as more services and values are considered, but the objective remains to produce the maximum possible value.

Managing for the production of desired ecosystem services is not without liabilities. Managing ecosystems solely on the basis of the monetary advantages they provide could promote the view that "nature is only worth conserving when it is, or can be made, profitable"

(McCauley 2006, 28). Given this pitfall, proponents must design tools and facilitate policy-making decisions with great care to ensure that the broadest reach of ecosystem functions and biodiversity is protected. Otherwise the initiative may result in minimum gains to selected services and potential harm to others (Daily and Matson 2008). In spite of these issues, an ecosystem-services approach provides a new model for coastal and ocean regions, which may presage a fundamental shift in coastal governance. If effectively implemented, this new approach could provide significant gains in sustainability.

Summary

The coast, a region of land and sea, is home to rapidly increasing human populations. Higher standards of living for coastal citizens frequently result in greater environmental impacts, while at the same time many in the populace increasingly value healthy, natural waters. As a result, coastal management professionals have confronted sustainability issues for four decades or more. In the Coastal Zone Management Act of 1972 administrators are challenged to both protect and develop the coast through the goals statement in the legislation. The 1987 Brundtland Commission report added a time dimension to the challenge and, while seeking sustainability of social and natural systems, noted the importance of managing the present in a way that does not jeopardize the future. More recently, coastal management analysis and action have been informed by ecosystem principles, which provide a means to shape the discussion about protect-develop-sustain within the context of environmental limits and human needs. In coming years coastal management will continue to test the ability of society to create and implement new governance systems matched to the desire for a sustainable future for coasts.

Richard BURROUGHS
University of Rhode Island

See also in the *Berkshire Encyclopedia of Sustainability* Best Management Practices (BMP); Catchment Management; Ecosystem Services; Edge Effects; Extreme Episodic Events; Food Webs; Global Climate Change; Large Marine Ecosystem (LME) Management and Assessment; Marine Protected Areas (MPAs); Ocean Resource Management

FURTHER READING

Birkland, Thomas. (2005). *An introduction to the policy process: Theories, concepts, and models of public policy making.* Armonk, NY: M. E. Sharpe.

Brauman, Kate; Daily, Gretchen; Duarte, T. Ka'eo; & Mooney, Harold. (2007). The nature and value of ecosystem services: An overview highlighting hydrologic services. *Annual Review of Environment and Resources, 32,* 67–98.

Burroughs, Richard. (1981). OCS oil and gas: Relationships between resource management and environmental research. *Coastal Zone Management Journal, 9,* 77–88.

Burroughs, Richard. (2011). *Coastal governance.* Washington, DC: Island Press.

Burroughs, Richard. (2012). Sustainability trajectories for urban waters. In M. Weinstein & R. E. Turner (Eds.), *Sustainability science: The emerging paradigm and the urban environment.* New York: Springer.

Clark, Tim. (2002). *The policy process: A practical guide for natural resource professionals.* New Haven, CT: Yale University Press.

Cortner, Hanna, & Moote, Margaret. (1999). A paradigm shift? In Hanna J. Cortner & Margaret Moote, *The politics of ecosystem management* (pp. 37–55). Washington, DC: Island Press.

Costanza, Robert, et al. (1998). Principles for sustainable governance of the oceans. *Science, 281*(5374), 198–199.

Crossett, Kristen; Culliton, Thomas; Wiley, Peter; & Goodspeed, Timothy. (2004). *Population trends along the coastal United States, 1980–2008.* Washington, DC: National Oceanic and Atmospheric Administration.

Crossland, Christopher; Kremer, Hertwig; Lindeboom, Han; Crossland, Janet; & Le Tissier, Martin. (Eds.). (2005). *Coastal fluxes in the Anthropocene.* Berlin: Springer-Verlag.

Day, John; Hall, Charles; Kemp, W. Michael; & Yanez-Arancibia, Alejandro. (1989). *Estuarine ecology.* New York: John Wiley & Sons.

Daily, Gretchen. (1997). Introduction: What are ecosystem services? In Gretchen Daily (Ed.), *Nature's services: Societal dependence on natural ecosystems* (pp. 1–10). Washington, DC: Island Press.

Daily, Gretchen, & Matson, Pamela. (2008). Ecosystem services: From theory to implementation. *Proceedings of the National Academy of Sciences, 105*(28), 9455–9456.

Fischer, Joern, et al. (2007). Mind the sustainability gap. *TRENDS in Ecology and Evolution, 22*(12), 621–624.

Heal, Geoffery, et al. (2001). Protecting natural capital through ecosystem service districts. *Stanford Environmental Law Journal, 20,* 333–364.

Juda, Lawrence. (1993). Ocean policy, multi-use management, and the cumulative impact of piecemeal change: The case of the United States outer continental shelf. *Ocean Development and International Law, 24,* 355–376.

Lant, Christopher; Ruhl, J. B.; & Kraft, Steven. (2008). The tragedy of ecosystem services. *Bioscience, 58*(10), 969–974.

Lasswell, Harold. (1971). *A Pre-View of Policy Sciences.* New York: American Elsevier.

Layzer, Judith. (2008). *Natural experiments: Ecosystem-based management and the environment.* Cambridge, MA: The MIT Press.

Levin, Simon, & Lubchenco, Jane. (2008). Resilience, robustness, and marine ecosystem-based management. *Bioscience, 58*(1), 27–32.

Lindblom, Charles. (1959). The science of "muddling through." *Public Administration Review, 19*(2), 79–88.

Lotze, Heike, et al. (2006). Depletion, degradation, and recovery potential of estuaries and coastal seas. *Science, 312,* 1806–1809.

McCauley, Douglas. (2006). Selling out on nature. *Nature, 443*(7), 27–28.

McLeod, Karen, & Leslie, Heather. (2009). State of the practice. In K. McLeod & H. Leslie (Eds.), *Ecosystem-based management for the oceans* (pp. 314–321). Washington, DC: Island Press.

National Research Council (NRC). (1999). *Our common journey.* Washington, DC: National Academy Press.

Pew Oceans Commission. (2003). *America's living oceans: Charting a course for sea change.* Arlington, VA: Pew Oceans Commission.

Ruhl, J. B.; Kraft, Steven; & Lant, C. (2007). *The law and policy of ecosystem services*. Washington, DC: Island Press.

Slocombe, D. Scott. (1998). Defining goals and criteria for ecosystem-based management. *Environmental Management*, *22*(4), 483–493.

United Nations Environment Programme. (2010). *Blue harvest: Inland fisheries as an ecosystem service*. WorldFish Center, Penang, Malaysia: UNEP.

United Nations Environment Programme. (2006). *Marine and coastal ecosystems and human well-being: A synthesis report based on the findings of the Millennium Ecosystem Assessment*. Nairobi, Kenya: UNEP.

US Commission on Ocean Policy (USCOP). (2004). *An ocean blueprint for the 21st century: Final report*. Retrieved December 13, 2011, from http://www.oceancommission.gov/documents/full_color_rpt/welcome.html

World Commission on Environment and Development (WCED). (1987). *Our common future*. New York: Oxford University Press.

Yaffee, Steven. (1999). Three faces of ecosystem management. *Conservation Biology*, *13*(4), 713–725.

Ecosystem Health Indicators

Improving the health of ecosystems is a long-term goal globally. Achieving this depends importantly on the establishment of meaningful ecosystem health indicators to identify critical drivers and pressures of ecological degradation and assess progress in addressing them. Indicators of ecosystem health and assessments based on them are essential to motivate both policy makers and the public to take the necessary actions to restore the health of the world's ecosystems.

Over the past half century, the concept of "health," long applied in reference to the vitality of individuals (humans as well as other species), has been extended to the higher levels of biological organization: populations, ecological communities, whole ecosystems, landscapes, and the biosphere. At each level, the metrics required to assess health differ. For example, the health of a population is not simply the summation of the health status of the individuals. New metrics are used that refer specifically to the status of an entire population. As for what constitutes ecosystem health, while there are a variety of definitions, nearly all of these comprise three basic elements: organization, vitality, and resilience. *Organization* refers to the structure of the ecosystem—the web of connections that link species with one another and with their environment. Loss of key structural properties of ecosystems—for example, disturbance to or removal of soil, coral, or stream bed—may cripple the ecosystem's capacity to maintain its biotic assemblage. *Vitality* refers to the overall metabolism of the ecosystem, that is, its capacity to maintain the flow of energy from primary producers (plants) to primary (herbivores) and secondary (carnivores) consumers, as well as its capacity to sustain the essential nutrient cycles. *Resilience* refers to the capacity of an ecosystem to recover from perturbations such as those caused by floods, fire, insect infestations, drought, and the like. While such perturbations cause short-term disruptions in community structure and ecological functions (drought, for example, may completely eliminate above-ground biota), healthy ecosystems are able to bounce back from these natural disturbances. For some perturbation-dependent ecosystems, natural disturbances such as forest fires or floods are an essential part of the dynamics to maintain the health of the system.

There is an ever-expanding list of metrics that are used to evaluate the health of ecosystems. Many of these are useful across the full range of the world's ecosystems. No single metric is adequate to evaluate health; rather, a suite of well-chosen indicators are employed. For indicators to be of service, it is necessary to establish the normal range of values for each indicator for the particular type of ecosystem under evaluation, for example, the normal range of primary productivity for grassland in a particular region, or the range of diversity of avian species in an old-growth temperate rain forest. Common indicators for a wide range of ecosystem health assessments include progressive dominance by opportunistic species, progressive invasion of nonlocal or non-native species, shifts in community structure, loss of substratum, disruption of nutrient cycling, and progressive loss of ecosystem services (attributes valued by humans) (Rapport and Whitford 1999, 193–203).

From Indicators to Indices

In descriptions of the overall condition of systems with large numbers of components, it is tempting to look for possibilities to create indices that aggregate information from individual indicators. Indices are valuable when they have a strong logical and scientific basis that allows for the amalgamation of disparate sets of data. This has been successful in certain social applications, such as the widespread use of a consumer price index, where the interpretation of the index is clear. Indices have been created for a variety of environmental applications (e.g., water quality, air pollution). In evaluating the health of ecosystems, there have also been attempts to amalgamate various considerations into a single index that would convey a general message on the health of the system. An example is the construction of a Forest Capital Index (Rapport and Ullsten 2006, 268–290). Such indices may give a crude overview of the health of a system, but they are often problematic as tools of communication. They are difficult to conceptualize if they combine many indicators related to different aspects of ecosystem health and the weighting (explicit or implicit) of the indicators that form the index heavily influences the message that the index conveys.

History of Indicator Development

The roots of indicator development are embedded in the history and prehistory of human culture. A remark attributed to Plato suggests that thousands of years ago people understood that certain modifications of agricultural drainage systems had adverse consequences for agricultural yields. Throughout the ages keen observers have taken note of correlations between human activity and ecosystem transformation. In the late seventeenth and early eighteenth centuries, when the rivers Thames and Rhine came under heavy stress from industrial waste, it was easy to recognize that foul smells, discolored waters, and local fish kills were a direct consequence of industrial activity.

In the twentieth century, the observations of naturalists, most notably those of Aldo Leopold in the 1940s, led to the realization that the consequences of human actions were rendering the land dysfunctional—resulting in what Leopold termed "land sickness." Among the worrisome signs Leopold observed in his native Wisconsin landscape were losses of native species; declines in biodiversity, soil fertility, and crop yields; reductions in biological productivity; increased presence of invasive species; and increased prevalence of diseases in both plants and animals. Several decades later, statistical agencies began to include the environment as part of their reporting functions, seeking a comprehensive framework that would relate human activities to environmental change. Statistics Canada's Stress-Response model provided a suitable template (Rapport and Friend 1979). Quickly adopted by the environmental secretariat of the Organization for Economic Cooperation and Development (OECD), this model, known today as the Pressure-State-Response (PSR) model, provides a taxonomy of anthropogenic stresses, ecosystem health indicators, and policy responses. Slightly modified, it is extensively used by the European Environment Agency for its state of environment reports.

A notable example of the application of the PSR model to assess ecosystem health is the retrospective assessment of the Baltic Sea undertaken by the Helsinki Commission (HELCOM 2010a). This report, *Ecosystem Health of the Baltic Sea*, is exemplary in showing the practicality of obtaining a suite of quantitative indicators of the health of a large-scale ecosystem and relating the state of health of this ecosystem to the suite of anthropogenic stresses impacting it. The value of this and many similar exercises—for example, evaluations of the health of the Laurentian Great Lakes (United States/Canada), Moreton Bay (Australia), Murray-Darling Basin (Australia), Bay of Fundy (Canada), Mesoamerican coral reefs, Florida Everglades—lies in providing a sound scientific basis for public policy geared to ameliorating environmental degradation.

Scale of Application

Indicators of the health of ecosystems may apply broadly across many ecosystems—or be more narrowly restricted to the characteristics of a particular ecosystem. For example, the prevalence of a particular forest insect pest, the mountain pine beetle, is a key indicator of the health of the coniferous forests of northern and central British Columbia, where an unprecedented outbreak (attributable to warmer winters and even age stands) has resulted in catastrophic loss of the forests of this region of Canada. This species, however, is specific to a particular ecological zone (mainly forests dominated by lodgepole pine, *Pinus contorta*). A more general indicator applicable to all forest ecosystems would be forest insect pest prevalence, without focusing on a specific species. Scaling up further, one could look at disease prevalence within an ecosystem. At this scale, the indicator could apply not only to forests but also to grasslands, lakes, marshlands, and so forth.

Scale is important in assessing ecosystem health. For example, monitoring the impacts of industrial sites is often limited to assessing the presence or absence of indicator species or potentially toxic compounds found in local biota. Such information may have little relevance beyond a very local domain. But when the pollution involves long-lived toxic substances that bio-accumulate in the food web, the consequences can eventually be reflected in ecosystem-wide degradation. Another example of widespread ecosystem-level impacts from industrial pollution is acidification of lakes and streams. In the 1970s, the sulfur dioxide emissions from a smelter near Sudbury, Ontario, were responsible for transforming a once-healthy mixed deciduous/coniferous forest to a virtual moonscape, eliminating forest cover and most vegetation on some 46,000 hectares, and in addition causing the acidification of more than seven thousand lakes in the region.

Quantitative Indicators for Ecosystem Health

An indicator is more than just the actual variable for which data is available. Massive algal blooms in a lake or a coastal area do not simply tell us that there are plenty of algae in the water; they also indicate that excessive nutrients are entering the water body due to either poor sewage treatment or unsustainable farming practices that waste nutrients. In a similar way, changes in bird populations can be linked with transformations in forests: the decline of species associated with old-growth forest indicates losses of particular habitats.

Why would one want to quantify the health of an ecosystem? Quantified or quantifiable measures of health and health change make it possible to evaluate actions that may improve or degrade health. By tracking the quantitative health estimate together with quantitative information on anthropogenic stresses, one may eventually be able to provide prescriptive information on what to do or not to do. For example, what intensity of fishing is compatible with sustaining the health of a lake or a sea area? What kinds of forestry practices are consistent with maintaining a healthy forest (that is, not just sustaining timber extraction)?

Traditional resource management has tried to approach these questions by examining each single exploitable resource as an independent entity for which it is possible to determine a maximum sustainable yield. This approach has proven to be inadequate in that, by focusing only on the resource, it neglects the interactions between different components of the ecosystem and thus falls short of assessing the overall health of the ecosystem. The maximum sustainable yield for a single species may actually be disastrous for the health of the whole ecosystem. Intense exploitation of forage fish may cause bird populations to decline; sustainable-yield harvesting of timber can lead to the demise of species dependent on dead wood.

A suite of quantitative indicators is required in order to cover the many facets of ecosystem health. For example, for the HELCOM retrospective assessment of the Baltic Sea, quantitative estimates of nutrient status, contaminant levels, state of fish stocks, and biodiversity were required. This allowed for the assessment of the overall ecosystem health of the Baltic, as well as the health of its sub-basins and regions. The value of such assessments lies in identifying critical issues that need immediate attention if the health of the ecosystem (upon which human well-being vitally depends) is to be improved. Similar approaches have been taken in North America's Laurentian Great Lakes, and yearly reports are published on a suite of ecosystem health indicators for each of the lake basins. These indicators and assessments provide guidance for both management and policy.

Monitoring Ecosystem Health

Monitoring the health of ecosystems requires not only careful selection of a suite of key indicators, but also attention to the practicality of gathering high-quality, reliable data on their status and trends, and the establishment of valid baselines (or normal ranges) in order to assess the health of the ecosystem. In some instances, this is relatively straightforward. For example, much can be learned about the health of the Baltic Sea from the abundance and distribution of an easily identified seaweed, bladderwrack (*Fucus vesiculosus*), found in the shallow waters of the Baltic Sea. The maximum depth at which this seaweed is found is limited by water transparency, and thus by the availability of light. Observing over time the depths at which this seaweed is found provides useful information on the overall eutrophic (nutrient) condition of the sea. Eutrophication (i.e., excess nutrients from sewage and agricultural runoff) is one of the main environmental issues in the Baltic. Monitoring for the bladderwrack serves double duty in that the seaweed provides habitat for many species, including many juvenile fishes. As its abundance increases, fish habitat improves, contributing to an improvement in the health of the entire ecosystem.

The baseline or reference level for each indicator is a crucial aspect of monitoring for ecosystem health. When it comes to anthropogenic synthetic pollutants such as polychlorinated biphenyls (PCBs), one can use zero as a reference level, but in addition one would also need to know the thresholds after which adverse effects begin to appear. The classical approach is to experimentally determine a dose-response curve; "no effect concentrations" can be estimated from the curve. These thresholds may provide guidance for monitoring ecosystem health, but one should be aware that the values commonly refer to laboratory tests using single substances, whereas ecosystems are subject to cumulative effects of multiple stressors. Thus the threshold levels for adverse effect may be significantly lower than in laboratory experiments. In some cases, different stressors may have antagonistic effects, and then the actual thresholds for adverse effects on the ecosystem may be higher than those anticipated in laboratory studies.

The baseline for variables such as species abundance or size distribution is more difficult to determine. In some cases it may be possible to derive an estimate of the conditions of the ecosystem when it was in a pristine state before significant human intervention. With the exception of some remote ecosystems, this approach is not practical, however, because many ecosystems have evolved in constant interaction with humans. Owing to such difficulties, often more pragmatic approaches are called into play in evaluating the health of ecosystems. In the European Union (EU), the Water Framework Directive and the Marine Strategy Directive (both legal documents) have created obligations to specify what constitutes "good ecological status" for different water bodies. This has led to identification of variables and thresholds that relate to ecological status (which in principle is closely related to health throughout Europe). The goal of these directives is to stimulate action plans that will result in restoring all water bodies to a healthy condition.

Trade-offs and Challenges in Ecological Monitoring

One of the major challenges in the development of ecosystem health indicators is finding indicators that can serve as "early warning" signals for ecosystem degradation. Too often these indicators are only discovered after the damage is done, and it is too late or economically impractical to reverse course. Generally, it is only in retrospect that changes in a seemingly insignificant part of the ecosystem might be linked to major ecosystem-wide consequences. For example, in the 1950s, mayflies (*Hexagenia* spp.), which became very abundant in spring breeding swarms along Lake Erie shores (United States/Canada), suddenly disappeared. While their disappearance did not go unnoticed (their swarming was always a nuisance to residents and resulted in slippery and thus hazardous driving conditions), the more far-reaching consequences for the health of Lake Erie were not immediately foreseen. Yet, in retrospect, the disappearance of mayfly larvae was one of the first signs of degradation of Lake Erie—for the disappearance of the larvae in the lake's bottom waters was due to nutrient loading and the creation of a seasonal dead zone in these bottom waters. In other situations, for example in lakes impacted by acidification, the earliest indicator that the level of acidification is negatively affecting the health of the lake might be found in altered behaviors of sensitive species. Yet, monitoring for behavioral changes in individuals or populations is extremely costly and generally impractical.

Another major challenge is the cost of acquiring synoptic data. With the rapid development of remote sensing imagery, and ever more numerous applications to ecosystem assessment, these costs are rapidly declining. Remote sensing is particularly well suited to spatial data on the extent of forest cover, forest tree composition, extent and location of wetlands, condition of grasslands, including surrogate measures for primary production, large-scale algal blooms, shoreline degradation, and the like. Yet for many variables that are essential in the evaluation of the health of ecosystems (such as fish stocks, water chemistry, soil biota, invasive species, disease prevalence, and so forth), remote sensing is of limited service. Intensive and often costly on-the-ground sampling is still required. In such cases, budgets require trade-offs among the number of variables being monitored, the frequency of data collection, and the intensity of the sampling design. All too often, such sampling results for large-scale ecosystems are far from ideal. Careful planning is therefore needed to combine different types of data collection so as to produce reliable indicators. It is particularly demanding to capture transient events

that may reveal early stages of the degradation of the ecosystem health.

Outlook for the Coming Decades

The general goal to improve and secure the health of ecosystems has become widely accepted. For example, in the Helsinki Commission's first retrospective assessment of the state of the Baltic Sea, ecosystem health was the explicit focus. An emphasis on healthy ecosystems is also incorporated into the long term vision for the International Union for the Conservation of Nature, the strategy for 2009–2013 of the European Environment Agency, assessments of many large-scale ecosystems such as grasslands in Inner Mongolia, catchments in Australia, tropical reef ecosystems, and forest ecosystems. It has been implicitly and explicitly expressed for both aquatic and terrestrial ecosystems, and it has been included in legal documents. At the same time, it is obvious that this is a long-term goal. Many of the world's ecosystems display all of the signs of ecosystem distress syndrome, the result of decades—in some instances, centuries—of cumulative anthropogenic stress. In many cases, it is unlikely that the damage done can be quickly repaired. Yet these difficulties should by no means dissuade society from its efforts to improve the health of the world's ecosystems.

Ecosystem degradation has engulfed large areas of the planet. Researchers argue that we may already be well into the sixth mass extinction of life. Desertification has taken over once-productive grasslands; the world's oceans have been vastly depleted of their natural bounty of wild fish stocks; and whole ecosystems, including the world's tropical coral reefs (hot spots of biodiversity), have become endangered and are at risk of extinction within this century. The Aral Sea, once the world's fourth-largest lake, has practically disappeared and has left in its wake a highly toxic environment in which millions of people still reside. Many ecosystems such as the Great Lakes and the Baltic Sea remain highly degraded, despite decades of efforts to restore their health. Unless and until there is widespread recognition that the very foundations for sustaining human and other life on the planet are rapidly eroding, it seems unlikely that the will to change course will arise.

Indicators of ecosystem health are capable of providing a robust measuring rod for assessing in which direction we are heading. Such indicators are an essential ingredient for addressing the situation, as it is not possible to act responsibly without a clear picture of what the state of ecosystems actually is. The other indispensible ingredient, however, is the political and social will to act on the findings. Ecosystem health indicators and assessments based on

them can help policy makers and the public see the need to act to maintain and restore the health of ecosystems.

David J. RAPPORT
EcoHealth Consulting

Mikael HILDÉN
Finnish Environment Institute (SYKE)

See also in the *Berkshire Encyclopedia of Sustainability* Biological Indicators (*several articles*); Computer Modeling; Ecological Footprint Accounting; Fisheries Indicators, Freshwater; Fisheries Indicators, Marine; Genuine Progress Indicators (GPI); Global Environmental Outlook (GEO) Reports; Human Appropriation of Net Primary Production (HANPP); Index of Biological Integrity (IBI); Land-Use and Land-Cover Change; Ocean Acidification—Measurement; Remote Sensing; Systems Thinking

FURTHER READING

Doren, Robert F.; Trexler, Joel C.; Gottlieb, Andrew D.; & Harwell, Matthew C. (2009). Ecological indicators for system-wide assessment of the greater everglades ecosystem restoration program. *Ecological Indicators, 9*(6, 1), S2–S16.

Helsinki Commission (HELCOM). (2010a). *Ecosystem health of the Baltic Sea*. Retrieved September 26, 2011, from http://www.helcom.fi/stc/files/Publications/Proceedings/bsep122.pdf

Helsinki Commission (HELCOM). (2010b). Homepage. Retrieved April 30, 2011, from http://www.helcom.fi/

Herrera-Silveira, Jorge A., & Morales-Ojeda, Sara M. (2009). Evaluation of the health status of a coastal ecosystem in southeast Mexico: Assessment of water quality, phytoplankton and submerged aquatic vegetation. *Marine Pollution Bulletin, 59*(1–3), 72–86.

Hildén, Mikael, & Rosenström, Ulla. (2008). The use of indicators for sustainable development. *Sustainable Development, 16*(4), 237–240.

Rapport, David J. (2007). Sustainability science: An ecohealth perspective. *Sustainability Science, 2*(1), 77–84.

Rapport, David J. (2010). How healthy are our ecosystems? In Bruce Mitchell (Ed.), *Resource and environmental management in Canada: Addressing conflict and uncertainty* (4th ed., pp. 69–96). Toronto: Oxford University Press.

Rapport, David J., & Friend, Anthony. (1979). *Towards a comprehensive framework for environmental statistics: A stress-response approach* (Statistics Canada Catalogue 11-510). Ottawa, Canada: Minister of Supply and Services Canada.

Rapport, David J., & Maffi, Luisa. (2011). Eco-cultural health, global health and sustainability. *Ecological Research, 26*(6),1039–1049.

Rapport, David J., & Singh, Ashbindu. (2006). An ecohealth based framework for state of environment reporting. *Ecological Indicators, 6*(2), 409–428.

Rapport, David J., & Ullsten, Ola. (2006). Managing for sustainability: Ecological footprints, ecosystem health and the forest capital index. In Philip Lawn (Ed.), *Sustainable development indicators in ecological economics* (pp. 268–290). Northampton, MA: Edward Elgar.

Rapport, David J., & Whitford, Walter G. (1999). How ecosystems respond to stress. *BioScience, 49*(3), 193–203.

Rapport, David J., et al. (Eds.). (2003). *Managing for healthy ecosystems*. Boca Raton, FL: CRC Press.

Wiegand, Jessica; Raffaelli, Dave; Smart, James C. R.; & White, Piran C. L. (2010). Assessment of temporal trends in ecosystem health using an holistic indicator. *Journal of Environmental Management, 91*(7), 1446–1455.

Externality Valuation

Externality valuation is a method that assesses social costs caused by production or consumption activities. It is used for project and policy appraisal but also as a cornerstone of green national accounting, in which environmental assets are factored into national wealth. The concept of ecosystem services in conservation has enriched our understanding of externality valuation, promoting research on the appropriateness of monetary valuation for ecosystem sustainability performance.

An externality is a state of affairs in which the action of a person affects (positively or negatively) the production or consumption possibilities of another, without this action being mirrored in a corresponding financial compensation. The most widely publicized externalities of interest with regard to sustainable environments are of a negative nature: the smokestacks of industrial plants emitting smoke that reduces visibility and leads to cardiovascular diseases. These and many other externalities causing pollution, resource depletion, and global change are pervasive and subtle phenomena of today's societies. Externality valuation can be a useful approach in the consideration of environment and sustainability issues through the application of monetary values to economic benefits or losses deriving from changes in the natural environment.

By generating negative side effects that affect third parties but are not appropriately taken care of by its originator, an externality gives rise to social costs, such as the decrease in well-being caused to those affected by reduced visibility and health in the foregoing example. When production and consumption activities cause extensive social costs, the economy is not allocating its resources to the society's best advantage. The ensuing divergence between private and social cost is taken care of by the process of internalization of social cost—that

is, the adding-up of social and private cost components. Externality valuation is the identification of and assignment of monetary values to social costs and thus is the process by which monetary values are attached to the physical impacts of externalities.

Origins

Externality valuation originated in the efforts of the United States Corps of Engineers to manage navigation projects in the beginning of the twentieth century. In that early phase, the administrative bodies involved were explicitly seeking to quantify and value externalities "to whomsoever they may accrue" (Flood Control Act of 1936). The slowly emerging practice of externality valuation was in search of a solid analytical structure when, in the 1950s, specific methodologies for the valuation of nonmarket externalities began to emerge. The US economist Harold Hotelling proposed capturing nonmarket benefits of national parks in the United States through what has become known as the *travel cost method.* This method used direct and indirect expenses of visitors to a national park as a proxy for the value of the park. The approach has spawned the literature on revealed preferences that has dominated the estimation of nonmarket values for recreation sites in general (Shaw 2005).

Externality valuation received a methodological turn in the 1960s when the US economist John Krutilla introduced the concept of nonuse value. In his own words: "There are many persons who obtain satisfaction from mere knowledge that part of wilderness North America remains even though they would be appalled by the prospect of being exposed to it" (Krutilla 1967, 781). Krutilla's idea to disassociate the economic value of an object from its use has rendered revealed preference methods, such as travel cost, inadequate to capture the total economic

45

value of natural environments because by definition non-use values leave no behavioral traces to evaluate. Stated preferences have emerged in the 1970s as an alternative, promising avenue of an inclusive externality valuation (Smith 2000). Stated preferences approaches include mainly choice experiments (CE) and contingent valuation (CV), both survey techniques making use of structured questionnaires to elicit individual welfare measures of changes in the provision of environmental goods and services (Bateman et al. 2002). The CE method has been practiced widely in the marketing domain since the late 1950s, where it is known as conjoint analysis (Green, Krieger, and Wind 2001). In the early 1990s, CV gained world prominence when used in the litigation procedure that the State of Alaska and the US government filed against Exxon for the nonuse damages caused by the *Exxon Valdez* oil spill (Carson et al. 2003).

Besides ecosystem damages, another domain where externality valuation has been practiced is energy planning. Concern about negative environmental effects of electricity-generating technologies has prompted a number of externality valuation studies in the United States (OTA 1994), which have replaced the previously dominant analytical approach of comparative risk assessment of alternative electricity supply options (Stirling 1997). Externality valuation studies in the energy domain proliferated in response to the adoption of environmental adders (environmental cost values added onto the monetary costs of energy generation) in the 1990s by US public utility commissions aiming at internalizing external cost in the investment and pricing decisions of electricity firms (Harrison and Nichols 1996).

Applications

In spite of several applications to date in the United States and an ever-expanding literature worldwide, the impact of externality valuation on efforts to quantifiably measure sustainability remains limited. Nonetheless, externality valuation has played a role in quantifying sustainability on three different levels of policy making. Firstly, on the level of individual project appraisal, assessing the full cost of alternative investment options has been applied, for example, in the case of environmental adders in the US energy sector, or water-related projects as defined in the European Union's Water Framework Directive initiated in 2000. Natural resource damage assessment in the United States, however, has tended less toward externality valuation than toward promoting a physical (service-to-service) assessment of lost ecosystem services (e.g., habitat equivalency analysis) and only in special cases a monetary (service-to-value) valuation of damages (Jones 2000).

Secondly, on the level of policy assessment, externality valuation has been used to assess monetary cost and benefits of past and proposed environmental legislation. A prominent example is efforts by the US Environmental Protection Agency (EPA) to estimate the cost and benefits of the Clean Air Act of 1970 and its amendments (Yang et al. 2004). A similar approach for Europe has been presented by a group of Norwegian environmentalists (Tollefsen et al. 2009) and the ExternE project (Bickel and Rainer 2005). A growing domain of externality valuation is climate economics, where the global externalities caused by climate change are increasingly given monetary value in order to substantiate what has come to be called "the social cost of carbon" (Tol 2008).

Thirdly, externality valuation has been used to assess macroeconomic sustainability performance by providing national economic accounts with monetary figures for depreciation of natural capital (Nordhaus 2006).

Controversies and Debates

Externality valuation is the locus of both opportunity and controversy for conservation specialists. In the limited cases where externalities can be valued on the basis of market prices the analysis is straightforward. By their very nature, though, externalities are nonmarket, public-good phenomena, demanding the skills of both ecologists and economists to evaluate. Major debates today cluster around both the broad question of the appropriateness of individualistic, economic theory as analytical background of externality valuation as well as narrower technical questions on the robustness of specific nonmarket valuation approaches.

Referring to the former, it has been argued that individuals face trade-offs in uses of the natural environment not as self-centered consumers seeking to maximize private benefits but rather as pro-social citizens aiming at achieving broader, altruistic concerns (Sagoff 1988). Even if the citizens–consumers divide is ignored, though, the usefulness of economic externality valuation is contested on the basis of the complexity of the natural environment, the resultant unfamiliarity with the goods and services to be valued, and consequently the nonexistence of well-defined preferences. In that case, what is actually happening is not the elicitation of pre-existing values but rather the construction of new, artificial preferences through the elicitation mechanism (Gregory, Lichtenstein, and Slovic 1993).

Referring to technical questions regarding theory of choice, practitioners emphasize aspects such as preference reversal (Tversky, Slovic, and Kahneman 1990), the hypothetical and nonbinding nature of values elicited through questionnaire surveys (Murphy et al. 2005), the sensitivity of value estimates to the valuation format (Cameron et al. 2002), and the analytical rigor and policy usefulness of nonuse values (Kopp 1992).

Sustainability Performance

Research on sustainability is increasingly based on the concept of ecosystem services—flows of benefits accruing to humans through well-functioning ecosystems (MEA 2005). The consideration of ecosystem services in conservation and sustainability debates has enriched our understanding of externality valuation and boosted research on the appropriate links between nature and economy (Kontogianni, Luck, and Skourtos 2010).

Currently, three major areas of research are shaping future developments in the field: Firstly, a more realistic representation of individual motives and constraints is replacing the abstract notion of *homo economicus* (Shogren and Taylor 2008). Secondly, *spatial realism* through geographic information systems (GIS) techniques is infused into the description of externalities and affected populations (Boyd 2008). Thirdly, and as a direct consequence of the former trends, the method of transferring the externality valuation results from the original study site to further policy sites is refined (Navrud and Ready 2007).

Areti KONTOGIANNI
University of the Aegean

See also in the *Berkshire Encyclopedia of Sustainability* Business Reporting Methods; Carbon Footprint; Community and Stakeholder Input; Computer Modeling; Cost-Benefit Analysis; Environmental Justice Indicators; Geographic Information Systems (GIS); Human Appropriation of Net Primary Production (HANPP); *The Limits to Growth;* Risk Assessment; Strategic Environmental Assessment (SEA); Systems Thinking; Transdisciplinary Research; Triple Bottom Line

FURTHER READING

Bateman, Ian J., et al. (2002). *Economic valuation with stated preference techniques: A manual.* Cheltenham, UK: Edward Elgar Publishing.

Bickel, Peter, & Rainer, Friedrich. (Eds.). (2005). *ExternE. Externalities of energy. Methodology update.* Luxembourg: European Commission.

Boyd, James. (2008). Location, location, location: The geography of ecosystem services. *Resources, 170,* 11–15.

Cameron, Trudy A.; Poe, Gregory L.; Ethier, Robert G.; & Schulze, William D. (2002). Alternative non-market value-elicitation methods: Are the underlying preferences the same? *Journal of Environmental Economics and Management, 44,* 391–425.

Carson, Richard T., et al. (2003). Contingent valuation and lost passive use: Damages from the *Exxon Valdez* oil spill. *Environmental and Resource Economics, 25,* 257–286. Flood Control Act of 1936. 33 U.S.C. §701–709 (1936).

Green, Paul E.; Krieger, Abba M.; & Wind, Yoram. (2001). Thirty years of conjoint analysis: Reflections and prospects. *Interfaces, 31,* S56–S73.

Gregory, Robin; Lichtenstein, Sarah; & Slovic, Paul. (1993). Valuing environmental resources: A constructive approach. *Journal of Risk and Uncertainty, 7,* 177–197.

Hanemann, W. Michael. (2006). The economic conception of water. In Peter P. Rogers, M. Ramon Llamas & Luis Martinez-Cortina (Eds.), *Water crisis: Myth or reality?* (pp. 61–92). London: Taylor & Francis.

Harrison, David, & Nichols, Albert L. (1996). Environmental adders in the real world. *Resource and Energy Economics, 18,* 491–509.

Jones, Carol A. (2000). Economic valuation of resource injuries in natural resource liability suits. *Journal of Water Resource Planning and Management, 126,* 358–365.

Kontogianni, Areti; Luck, Garry W.; & Skourtos, Michalis. (2010). Valuing ecosystem services on the basis of service-providing units: A potential approach to address the 'endpoint problem' and improve stated preference methods. *Ecological Economics, 69,* 1479–1487.

Kopp, Raymond J. (1992). Why existence value should be used in cost-benefit analysis. *Journal of Policy Analysis and Management, 11,* 123–130.

Krupnick, Alan J., & Burtraw, Dallas. (1996). The social costs of electricity. Do the numbers add up? *Resource and Energy Economics, 18,* 423–466.

Krutilla, John. (1967). Conservation reconsidered. *American Economic Review, 57,* 777–786.

Millennium Ecosystem Assessment (MEA). (2005). *Ecosystems and human well-being: Synthesis.* Washington, DC: Island Press.

Murphy, James J.; Allen, P. Geoffrey; Stevens, Thomas H.; & Weatherhead, Darryl. (2005). A meta-analysis of hypothetical bias in stated preference valuation. *Environmental and Resource Economics, 30,* 313–325.

Navrud, Stale, & Ready, Richard. (Eds.). (2007). *Environmental value transfer: Issues and methods.* Dordrecht, The Netherlands: Springer.

Nordhaus, William D. (2006). Principles of national accounting for nonmarket accounts. In Dale W. Jorgenson, Dale W.; Landefeld, J. Steven; & Nordhaus, William D. (Eds.). *A new architecture for the US national accounts* (pp. 143–160). Chicago: University of Chicago Press.

Office of Technology Assessment (OTA). (1994). *Studies of the environmental costs of electricity.* Washington, DC: US Government Printing Office.

Sagoff, Mark. (1988). *The economy of the Earth: Philosophy, law, and the environment.* Cambridge, UK: Cambridge University Press.

Shaw, W. Douglas. (2005). The road less traveled: Revealed preference and using the travel cost model to value environmental changes. *CHOICES, 20*(3), 183–188.

Shogren, Jason S., & Taylor, Laura O. (2008). On behavioral-environmental economics. *Review of Environmental Economics and Policy, 2,* 26–44.

Smith, V. Kerry. (2000). *JEEM* and non-market valuation: 1974–1998. *Journal of Environmental Economics and Management, 39,* 351–374.

Stirling, Andrew. (1997). Limits to the value of external costs. *Energy Policy, 25*(3), 517–540.

Tol, Richard S. J. (2008). The social cost of carbon: Trends, outliers and catastrophes. *Economics e-journal, 2,* 1–22. Retrieved October 13, 2011, from http://www.economics-ejournal.org/ej-search?SearchableText=tol

Tollefsen, Petter; Rypdal, Kristin; Torvanger, Asbjørn; & Rive, Nathan. (2009). Air pollution policies in Europe: Efficiency gains from integrating climate damage effects with damage costs to health and crops. *Environmental Science & Policy, 12,* 870–881.

Tversky, Amos; Slovic, Paul; & Kahneman, Daniel. (1990). The causes of preference reversal. *American Economic Review, 80,* 204–217.

Yang, Trend; Matus, Kira; Paltsev, Sergey; & Reilly, John. (2004). *Economic benefits of air pollution regulation in the USA: An integrated approach.* Cambridge, MA: MIT Joint Program on the Science and Policy of Global Change.

Fish

Marine and freshwater environments face challenges as modern fish harvesting methods put pressure on wild fish stocks. Aquaculture, in which fish are bred in ponds or enclosures in the sea, is the fastest growing technology. Although it preserves wild stocks, it presents its own ecological risks. Different forms of regulation have been implemented to protect fish stocks and the environment with varying levels of success.

Fish, crustaceans, mollusks, and other aquatic organisms, including seaweed and algae, are important sources of sustenance, employment, and recreation for millions of people around the world. (Throughout this article, the term *fish* is used generically to refer to all living aquatic organisms in marine and freshwater environments.) Many stocks of fish are severely overfished and are further threatened by other human activities like coastal development and waterborne pollution. Sustainable human interaction with these resources requires better understanding of marine and freshwater ecosystems and a willingness to limit human use of ocean resources.

Most of the fish landed globally is consumed by people. (Not all fish that are caught are brought to land and sold; less valuable species that are accidentally harvested are usually discarded at sea and so are not included in landings data.) In 2008, 81 percent (115 million tons) of fish was used as food by human beings. Of the remainder, 14 percent (20.8 million tons) was converted into fish meal and fish oil, which are additives in animal feed. The rest (4 percent, or 6.2 million tons) was used as bait, in pharmaceuticals, and directly as feed in aquaculture and furrier operations (FAO 2011, 9). In addition, millions of anglers, scuba divers, and tourists enjoy recreational activities associated with marine or inland

aquatic species, including fish, crustaceans, and mammals.

Fisheries in History

People began catching and eating aquatic organisms in prehistoric times. Archeological evidence suggests that sedentary species such as oysters and clams were exploited first, followed by more mobile species as people developed nets and traps to catch them. Hooks were not introduced until about fifty thousand years ago when protohumans began making tools out of bone (Diamond 1999, 39). Since that time, fishing and fish consumption have expanded geographically, technologically, and gastronomically.

It is most likely that human use of aquatic resources started on inland waters, like rivers, ponds, and lakes, and in coastal areas, like beaches, marshes, and tide pools. These areas were easy to access and did not require boats or other vessels, which were not developed until about 4000 BCE. The earliest written record of commercial fishing activities was found in Sumer, between the Tigris and Euphrates rivers in modern-day Iraq. Archives dated around 2300 BCE detail employment of hundreds of fishers who were organized in guilds and used varying types of gears in different areas of the two rivers. Some guilds created artificial ponds and stocked them with valuable fish species (Royce 1987, 74). In Asia, aquaculture also started on inland waters, mainly in China, around 2000–1000 BCE (Rabanal 1988).

Even with aquaculture, inland waters could not supply growing human populations indefinitely. As vessel and navigation technologies improved, fishers increasingly exploited marine species. Freed from the coasts by ocean-going vessels, whalers were some of the first fishers to travel long distances to harvest marine organisms. In fact, whalers from Britain and Spain (Basques) were the first Europeans to discover the massive stocks of cod in the North Atlantic (Ellis 2003, 61). Preserved using salting and drying techniques, cod was a major trade good from the fourteenth through the nineteenth centuries. Along with herring and other key species, cod added to the wealth of many European nations and also encouraged settlers to move to North American colonies, like Newfoundland, which were closer to the best fishing grounds.

European fishers developed schooners to harvest cod. These large sailing vessels could carry a fleet of small boats called coracles that could be deployed daily to cover a wider area using hand-operated longlines, long ropes set periodically with baited hooks. Japanese fishers of tuna developed a similar technique around the same time as the Europeans. Later, Europeans would replace lines with nets, either trawls pulled through the water or purse seines (used to surround a school of fish and then close or "pursed" at the bottom), which worked well on schooling species like cod and herring.

During the Industrial Revolution, these basic fishing methods (longline, trawl, and purse seine) became mechanized, and motorized vehicles replaced sailing vessels. Processing for trade changed with the industrialization of canning. The composition of catches changed, too, as fishers could target schools of smaller fish such as tuna, sardines, and mackerel on the open ocean. By World War I the scope and the scale of marine fishing were extensive, and some stocks showed signs of overfishing, but the two World Wars and the Great Depression substantially diminished fishing pressure for several decades.

Fishing after World War II

Marine fishery production took off again after World War II. By this time, inland waters were mostly closed to commercial fishing in the United States and Europe, as well as Japan, but these three fishing powers dominated fishing on the high seas. With bigger boats and gear, fishers could harvest massive amounts of fish. Generally, they caught large, nonschooling species such as swordfish on longlines that, due to monofilament technology, could stretch up to one hundred kilometers and carry more than three thousand hooks. Using monofilament lines (made from plastic), fishers also revived an ancient practice, the use of drift nets (nets set adrift with buoys to entangle any animal too large to slip through the mesh). Combined with mechanical winches, monofilament also enabled the use of trawl and purse seine nets large enough to cover several football fields.

As boat size increased with gear size, fishers had to travel farther to fill these huge vessels. In 1954, the British introduced the first factory trawler—a fishing vessel with processing facilities on board. This 75-meter-long ship, named the *Fairtry*, weighed 2,800 tons. Rather than carry the fish back to port in tanks of cold brine (saltwater), the *Fairtry* carried machines and workers to clean the fish, quick freeze the fillets, and even extract cod liver oil. By the 1960s, factory vessels were also in use in purse seine fisheries. Most of these ships were about 25 meters long and could catch and process up to 3,000 tons of fish per day, destined for canneries or other secondary processing facilities on land.

By the 1970s, air freight revolutionized fisheries around the world by enabling fast shipment of quality fish to the sushi and sashimi consumers of Japan. Demanding high-quality fish, Japanese buyers would pay top-dollar for certain species. Atlantic bluefin tuna went from a "trash fish," or bycatch, in fisheries for cannery-grade tuna (mostly yellowfin or skipjack) to one of the most expensive species in the world once it could be transported quickly to sushi consumers. Despite sushi's growing popularity internationally, Japan still imports the vast majority of bluefin tuna.

Fishers also began using new navigation and fish-finding technologies in this period. Sonar and satellite imagery are now common even on fairly small commercial and recreational vessels. Factory vessels often carry helicopters that greatly increase their ability to search for schools of fish. In the 1980s fishers began using fish aggregating devices (FADs) to create artificial schools of tuna and similar species. These floating platforms attract fish as natural logs or debris rafts do, but are equipped with a global positioning beacon so that they can easily be located. (While it is not known for certain why some species congregate around floating objects, some marine biologists theorize that they provide protection from predators as well as providing food; according to this theory, FADs work because fish confuse them with rafts of seaweed or plankton.)

With all these new technologies, modern industrial fishing fleets are incredibly efficient at catching fish. At the beginning of the twenty-first century, China is the world's largest producer of marine and freshwater species. Its fleets range globally targeting diverse fish. Peru and Indonesia round out the top three harvesting countries. Peru's fleets mainly fish within the Peruvian Exclusive Economic Zone and primarily target Pacific anchovetta, one of the more prolific species in the ocean. Indonesian production is about 60 percent capture fishing in marine and inland waters and 40 percent aquaculture. In 2008, production and export of processed fish products were also dominated by China, along with Thailand and Vietnam. The United States, European Union members, and Japan are the biggest importers of fish and related products; the value of fisheries imports was globally $14.9 billion in 2008 (FAO 2011, 10).

Noncommercial Uses

Commercial fishing is not the only use of fish resources globally; artisanal fishers and recreational users are also significant users. Artisanal fishers are small-scale producers who fish for subsistence or to sell their catch at local markets. Approximately 41 percent of the global fishing fleet is made up of artisanal fishers, and as much as half of the global catch is produced by small-scale fleets (FAO 2011, 7, 9). Although the line between artisanal and small-scale commercial fishing is blurry, both are labor intensive (80 percent of jobs in the fisheries sector) and mainly harvest fish for direct human consumption. They provide much needed nutrition and income in some less developed countries.

Recreational fishing, scuba diving, and other types of marine and freshwater tourism are also economically important. Global numbers of recreational users are unknown but are thought to be large and increasing. Regional studies, such as those conducted by US economist Ram K. Shrestha and colleagues in 2002 and US economist David K. Loomis in 2005, estimate that recreational fishers spend hundreds of dollars per trip on inland and marine fishing, for total values per area in the millions of dollars. Estimates are similar for scuba diving and snorkeling, which are major recreational activities in many coastal and small-island areas. Whale watching, glass-bottom boats, and similar activities are also common attractions around the world.

Depletion and Disruption

Even though fish are a renewable resource, there are limits on the amount that humans can harvest. Commercial exploitation of fisheries resources is much greater now than in the past, and noncommercial uses are also on the rise. To cope with rising demand, aquaculture, particularly of popular species like shrimp and salmon, has grown, accounting for most of the increase in production. Capture production (harvesting wild marine and freshwater organisms) continues on both artisanal and industrial scales. Both types of fishing present sustainability issues. Common problems with capture fisheries fit into two categories: the tragedy of the commons and negative ecosystem effects. Aquaculture can also have negative impacts on ecosystems and, because fish are the major feed in many aquaculture operations, it contributes to common problems in capture production.

Tragedy of the Commons

The tragedy of the commons is the best-known problem in fisheries today. Introduced by the US ecologist Garrett Hardin (1915–2003) in an article of the same name that appeared in *Science* in 1968, the "tragedy" results from a combination of "open access" (anyone can harvest fish from the oceans and public waterways) and "rivalness" (if one person harvests a fish, it cannot be taken by someone else). Because of these attributes, fishers have no incentives to leave fish behind to replenish the population; why hold back when someone else can easily catch the fish left behind? This typically creates a "race for the fish," which results in overexploitation (more than the efficient amount of fish is captured) and overcapitalization (more than the efficient amounts of capital and labor are invested in the fishing industry). In other words, without some form of limit to access, capture fisheries display boom-bust characteristics and ultimately end up in either subsistence or collapse as too many fishers spend too much time and money pursuing too few fish.

Freshwater Fisheries Depletion

The same ease of access that led to early exploitation of fisheries in inland waters facilitated early depletion of freshwater species. Even the Sumerians recorded large fluctuations in catch size and economic recession among fishing guilds. Their rulers intervened to protect the fisheries and ensure a relatively stable supply of fish (Royce 1987, 75). Few data are available on global stocks of freshwater fish, but global landings are estimated at about 10 million tons per year (Welcomme et al. 2010, 2881). A survey of the literature suggests that most natural freshwater fisheries are heavily depleted and overcapitalized. From the Great Lakes in North America to Lake Victoria in Africa, freshwater fishing communities struggle because the stocks of fish are too small and the number of fishers is too large.

Marine Fisheries Depletion

Many marine capture fisheries also display attributes of the tragedy of the commons. The UN Food and Agriculture Organization (FAO) estimates that about 32 percent of commercially harvested marine species are either overexploited or heavily depleted. Because these species are caught faster than they can reproduce, harvests will decline in the long term unless catches are reduced in the short term to allow the stocks to rebuild. Furthermore, about 53 percent of these species are fully exploited, which means that catches cannot increase in the short term without reducing harvests in the long run. Only 15 percent of currently exploited marine species can support an increase in catch rates over the long term (FAO 2011, 8).

Many of the most economically valuable commercial species are the most overexploited. Atlantic bluefin tuna, which can sell for as much as $400,000 per

270–360 kilogram fish, is so severely depleted that it was nominated for protection under the Convention on International Trade in Endangered Species of Wild Fauna and Flora (CITES) in 2010 (Webster 2011). Other economically valuable and severely overexploited species include Patagonian or Antarctic toothfish (known as Chilean sea bass), orange roughy, sturgeon (for caviar), and many species of sharks. Sharks are valued for their fins, which are used to make shark-fin soup, a popular delicacy in Asia. Most of these species have slow reproduction rates and therefore relatively small populations. The same biological scarcity that adds to their economic value makes them prone to overexploitation.

Overcapitalization is also a major problem for marine capture fisheries. The known global fishing fleet consists of more than 4.3 million vessels, of which 59 percent are powered by engines (FAO 2011, 7). Many of these are factory ships, but megavessels like the *Fairtry* are no longer cost-effective—they cannot catch enough fish to cover costs of production at current prices. Smaller-scale commercial fishing communities can have similar problems, especially in developed countries where costs of production are high. In fact, fishing effort has generally shifted to developing countries as boat owners seek lower costs of production and less overfished stocks. For instance, much of the fleet that currently operates under a Chinese flag originated in Japan but moved to Taiwan and then to China in pursuit of lower operating costs.

Other boat owners choose to operate under flags of convenience or as what has come to be known as the "illegal, unregulated, and unreported (IUU)" fleet. Under international maritime law, a fishing vessel is required to fly the flag of its home country when fishing on the high seas. Pirate fishing vessels either purchase a flag from a country with lax enforcement (flag of convenience), counterfeit their credentials, or fish without a flag (IUU). These practices are particularly problematic for lucrative tuna stocks as well as fisheries for toothfish, orange roughy, and various sharks. IUU fishing may be equivalent to as much as half the reported catches in these fisheries and so is a major concern for international fishing fleets and the nations that harbor them. Furthermore, IUU fleets can decimate fisheries in coastal waters if states lack enforcement capacity, often driving subsistence fishers to other sectors of the economy. For instance, many of the infamous Somali pirates were once fishers but took up piracy when fish stocks were depleted by foreign fleets.

Ecosystem Disruption

The fish that are caught are an integral part of marine and freshwater ecosystems. Commercial fishing can disrupt ecosystem functions in several ways. First, many fishing methods, such as long-lining and drift nets, capture large amounts of bycatch along with harvests of target species. Data on bycatch are difficult to collect, but global discards (fish thrown back once caught because they were not as valuable as the target species) in marine commercial fisheries were estimated to be about 7.3 million tons annually from 1992 to 2001, just under 10 percent of landed harvests in the period (Kelleher 2005, 17).

Bycatch species may not be economically important, but they can be ecologically important and are often less prolific than targeted species. Endangered sea turtles were once caught and killed in trawl fisheries targeting shrimp, until fishers began to use turtle-excluding devices to prevent such bycatch. Many other species, like white marlin in the Atlantic and albatross in the Pacific, are still endangered as a result of accidental interactions with fish gear. In addition, some types of gear, like bottom trawlers, destroy important habitats by scraping the bottom of the sea, resulting in much higher levels of bycatch of plants and animal species than most other methods.

Human use also disturbs global fish stocks by adding and subtracting substantially from marine and freshwater food webs. Many lakes and streams were completely altered by first the removal of native stocks through overfishing and the subsequent introduction of alien species through restocking programs. Accidental introduction of alien species also can be destructive if that species outcompetes others. Asian carp and zebra mussels are examples of invasive species in the United States. Elsewhere, coral reef ecosystems were negatively affected by the removal of large predators including tunas, sharks, and rockfish. Without these carnivores, the numbers of small fish exploded around reefs, causing overgrazing and, eventually, the bleaching of the reef and the death of the entire system. These types of "trophic cascades" are difficult to monitor and predict but may be increasingly important as humans fish down the food chain, chasing smaller species once those at the top of the marine food web are overexploited.

Pollution and coastal development can amplify the effects of overfishing. Species that are reduced in numbers and biodiversity because of overexploitation do not have the resilience to cope with toxic events such as algal blooms associated with nutrient runoff from land-based agriculture. Freshwater systems are most exposed to these threats, followed by coastal and then open-ocean regions. Large "dead zones" exist near the mouths of most major rivers: these are areas that are devoid of oxygen because fertilizer runoff feeds algal blooms that consume all the oxygen and then die off. Plastic is also a large problem for aquatic organisms. Many species perceive plastic waste in the water as food or may be trapped and killed in larger segments of plastic debris. Coastal and freshwater development contributes to pollution and destroys habitat through resurfacing and erosion. All of these factors negatively affect the resilience of fisheries resources.

Coral reefs and other aquatic ecosystems will also be affected by climate change. While some species may benefit—particularly tropical fish with high levels of mobility—many will not be able to adapt. There are two factors at play. First, because there is more carbon in the atmosphere, more is captured by the ocean, which increases its acidity. This could prove fatal to corals and other species that rely on hard shells made of calcium to survive. Second, as the ocean warms up, fish will need to move to stay within their preferred temperature zone. Fish that cannot move easily or are highly sensitive to temperature shifts will probably decline. Therefore, sedentary species like rockfish or Arctic and Antarctic species that cannot go farther north or south to find cold water will be most affected. It is unclear what ecosystem effects will result from the loss of such species or the proliferation of more adaptable fishes. There is also concern over climate effects on arctic krill, which are key prey for many species ranging from penguins to whales.

Environmental Impacts of Aquaculture

Because it is based on private ownership, aquaculture avoids the tragedy of the commons problem but creates problems for other species or supporting resources and environments. Many of the most valuable species in aquaculture, including salmon and bluefin tuna (bluefin are actually captured and then fattened, rather than farmed, so commons problems also remain for this species), are top predators and must be fed many tons of smaller fish that usually come from capture fisheries. Vegetarian fish such as carp and tilapia do not require fish for feed but, like all aquaculture operations, do produce waste. This waste introduces substantial amounts of nutrients into environments, which can lead to eutrophication—when an influx of nutrients causes an explosion in microbial life that smothers larger species. Such environmental degradation is known as an externality, a cost of production that is not usually included in the price.

Three other externalities associated with aquaculture are disease resistance, parasites, and genetic biodiversity. Aquaculture production requires that many fish be kept in a relatively small space, either in ponds (freshwater) or in pens in the sea (marine). In either case, proximity increases the transmission of disease, and antibiotics are used to improve survival. Antibiotic use increases the probability of disease resistance, reducing the benefits of the antibiotic for treatment and prophylaxis. Crowded conditions also allow parasites to breed more readily. Parasites may not be a problem for the adult fish in the pens but can seriously affect juveniles of wild species nearby. Aquaculture fish are usually bred or genetically altered for rapid growth or other desirable traits. If genetically altered fish escape from aquaculture pens, they may

interfere with wild populations as interbreeding or competition reduces genetic diversity.

Solutions

Government regulation and community action can be effective in encouraging sustainable fisheries management. As Elinor Ostrom, winner of the Nobel Prize for Economics, points out in her work, groups of resource users can overcome the tragedy of the commons through collective action to limit harvests. The lobster gangs of Maine are a famous example of successful nongovernmental regulation of a commercial fishery. They operate in different bays throughout the state and limit catches through a complicated schedule that indicates who can fish when and where. The entire community monitors compliance with the agreed regime; enforcement can range from a stern warning to the destruction of traps or a hole in an interloper's boat.

Fisheries Management

Response to the tragedy of the commons often takes the form of management by local, national, or international regulatory bodies. Regulations include gear restrictions, such as the bans on drift nets to protect marine mammals and other common bycatch species, or, more usually, limits on the size of nets or lines, the size of boats, or the use of fish-finding technologies. Seasons and other time or area closures are also implemented, although limiting the period or area in which fishers can operate generally results in the concentration of effort and is usually inefficient. For instance, when a fishing season is limited to a month or so, fishers will work as much as they possibly can in this short period. They also invest in larger, faster boats than they would otherwise, generating both capital and labor inefficiencies (higher wages or overtime as well as reduced productivity and increased injury due to fatigue). Quotas are another common type of restriction. In a quota system, the regulator sets a limit on the harvest (usually on an annual basis), monitors catches, and closes the fishery once the quota is reached. Governments can also limit entry by requiring that fishers obtain a license to fish and by charging fees that increase the costs of production.

Domestic regulation can be successful, particularly at rebuilding overfished stocks. In many countries, however, fishers wield political clout that is disproportionate to their economic contribution. Therefore, regulations may fall short of scientific advice or even be counterproductive. When governments provide subsidies to failing fisheries, they encourage additional entrants, actually increasing overexploitation and overcapitalization. Furthermore, many countries lack either the will or the capacity to monitor and enforce regulation, which undermines their effectiveness.

Because enforcement is difficult and information is scarce, many economists favor tradable quota or permit systems, which can mimic privatization by creating markets for rights of access. In this approach, known as individual traceable quota (ITQ), the government sets a quota and shares of the quota are distributed to fishers, usually by an auction or distribution based on historic catch rates. Fishers can choose to harvest their full share or to trade their quota by selling it to someone else. A market for quota shares is established, which gives fishers incentives to ensure that the system is enforced. Successful ITQ systems exist in New Zealand and in Iceland, but the method is not without risks. If a quota is set too high, shares have no value and the market collapses.

According to the United Nations Law of the Sea, which entered into force in 1995, national governments are legally able to manage stocks within their exclusive economic zone (EEZ), but they must cooperate to manage fisheries that fall outside or straddle the EEZ boundary. Examples of such fisheries include those targeting salmon in the north Atlantic, fisheries for tuna and swordfish in all oceans, and fisheries for toothfish and orange roughy in Antarctic waters. To manage such fisheries, countries established regional fisheries management organizations (RFMOs), which provide data on these types of stocks and, in some cases, facilitate the negotiation of multilateral management measures. Time or area closures and national quotas are the most common measures adopted by RFMOs.

Unfortunately, monitoring and enforcement on the high seas is very difficult, and IUU fishing and the dilution of management measures due to conflicts over allocation of quota shares are significant challenges. RFMOs established programs that track trade in high-value fish by requiring that paperwork travel with the fish through each point of sale. Import documentation could be compared with reported catches to identify cheating. In response, IUU fleets began "fish laundering," by creating false documents and shipping through RFMO member countries. RFMOs also imposed sanctions on known flag of convenience countries, which pushed many vessels into the IUU fleets but may have reduced pirate fishing somewhat.

New Approaches

Because traditional approaches to fisheries management are not universally successful, environmental and recreational fishing interests have started pushing for alternative methods of conservation. These methods fall into two main categories: consumer-based ecosystem labeling and government-based ecosystem management. There is also the precautionary approach to fisheries management, which requires that regulators act on scientific advice even though there may be associated uncertainties.

Although it is part of international fisheries law, the precautionary approach has not been universally accepted, and implementation is still problematic in most cases.

Several advocacy groups hope to harness the power of concerned consumers by promoting sustainable labeling. The first instance of sustainable labeling occurred during the dolphin-tuna controversy of the 1980s. Because dolphins were known to be caught and killed in tuna fisheries in the eastern Pacific Ocean, US consumers boycotted canned tuna. This boycott led to a breakdown in RFMO negotiations for the area and economic losses for fishers from many countries. Finally, a group of purse seiners invented a gear modification that would allow dolphin to escape tuna nets, and countries negotiated a new RFMO that would monitor fishing activities to ensure that no dolphins were harmed in the process. This RFMO created the dolphin-safe-tuna label that is still in use. Other labels used include turtle-safe shrimp and "sustainable seafood."

Ecologists and others who worry about the ecosystem effects of commercial fishing believe that bycatch reduction is the first step in improved fisheries management. They propose that regulations should be modified to take nonhuman use of fisheries into consideration. This includes multispecies management, in which traditional fisheries science is modified to include trophic interactions. Thus, scientists would consider predator-prey relationships as well as stock size and reproductive rates in the models that they use to inform management decisions.

Marine protected areas are also management tools. These limited areas would be easier to police and, if chosen correctly, could improve stocks outside the area as fish are given space to reproduce and replenish the stock. This thinking is closely related to marine-spatial planning, which integrates management of all uses of an area based on ecological as well as regulatory considerations. Habitat rehabilitation is also important to this approach, particularly for coastal or anadromous species that migrate from freshwater to the open ocean.

Controversies

While it is agreed that many fisheries around the world are overexploited, the degree of the problem and the nature of the solution are disputed. This conflict can best be seen in an exchange of papers published from 2006 through 2009 in the journal *Science*. First, in a global study, the Canadian marine ecologist Boris Worm (2006) and colleagues found that commercial fishing was so detrimental to the ocean's ecosystems that if not reversed it would decimate fish production, impede water quality regulation, and slow recovery from perturbations (p. 787) In response, the US fisheries scientist Ray Hilborne

(2007) and others published letters in the same journal, acknowledging the problem of overfishing but also pointing out successful management of fisheries around the world. They also argued that the Worm et al. results were skewed because the data came from long-run fisheries—those most likely to be overfished. Subsequently, these teams collaborated, producing an article (Worm et al. 2009) that reconciled the two perspectives by combining datasets and focusing on local successes.

Because the effects of overfishing are very evident for many species, managers have tried to reduce fishing effort. This type of initiative leads to a second type of controversy: deciding who should stop fishing. Indeed, conflicts over access pre-date conflicts over science in the United States, where a legal battle over the right to harvest cod off Massachusetts initiated the first national marine science program in the 1870s. Today, regulators are often pressured to inflate quotas and catch limits to accommodate competing users. This pressure is a particular problem for international fisheries, since RFMOs are based on negotiation with no higher power of appeal. Conflicts over access at RFMOs most often arise between historical fishing countries (the United States, members of the European Union, and Japan) and developing fishing countries (including Brazil, Mexico, China, and Ghana).

Commercial and recreational interest groups often press for more stringent regulations than commercial fishers want. Some conservation groups believe that, because of the difficulties with monitoring and enforcement, the only way to manage world fisheries is to outlaw commercial fishing. Given that this same argument was used to successfully eliminate commercial whaling in the 1970s and 1980s, commercial fishers have reason to fear such claims. Many conservation-oriented groups are more moderate, however, and seek ways to improve both the biological and the economic health of global fisheries.

Outlook

The outlook for global fisheries is mixed. There is potential for sustainable management, but only if consumers, managers, and fishers accept the limits of the resource. Some species, such as Patagonian toothfish or Atlantic bluefin tuna, may be driven to extremely low levels, causing collapse and possibly extinction. It was previously believed that more prolific and lower-priced species will last longer and have a better chance to be sustainably managed, but there is recent evidence that suggests that these stocks are more likely to collapse due to combinations of overfishing and environmental factors. Indeed, Peruvian anchovetta, one of the most prolific stocks in the oceans, has collapsed several times in response to combined stressors (Pinsky et al. 2011).

Monitoring and enforcement capacity is increasing globally, and the understanding of fisheries science is also greatly improving. Nevertheless, there is considerable overcapacity in global fishing fleets, and this prevents sustainable management. Growing public awareness may produce sufficient political will to generate better management. For international fisheries, rising oil prices may force some reduction in fishing, since higher fuel costs fall heaviest on distant water fishers. Technological fixes can also reduce bycatch in many fisheries, and management measures are working at the local scale in many countries. Aquaculture, too, is has potential benefits, as long as it is regulated to limit environmental degradation and ecosystem effects. Ultimately, the sustainability of the oceans is tied to the sustainability of the planet as a whole, since both population and affluence drive demand for fish and therefore accommodate overfishing.

D. G. WEBSTER
Dartmouth College

See also in the *Berkshire Encyclopedia of Sustainability* Dams and Reservoirs; Food, Frozen; Oceans and Seas; Parks and Preserves—Marine; Rivers; Water (Overview); Water Energy

FURTHER READING

Diamond, Jared. (1999). *Guns, germs, and steel*. New York: W. W. Norton & Company.

Ellis, Richard. (2003). *The empty ocean*. Washington, DC: Island Press.

Food and Agriculture Organization of the United Nations (FAO). (2011). The state of world fisheries and aquaculture 2010. Rome: FAO.

Hilborn, Ray. (2007, June 1). Biodiversity loss in the ocean: How bad is it? *Science, 316*, 1281–1282.

Kelleher, Kieran. (2005). Discards in the world's marine fisheries: An update (FAO Fisheries Technical Paper 470). Rome: Food and Agricultural Organization of the United Nations.

Loomis, John. (2005). The economic value of recreational fishing & boating to visitors & communities along the Upper Snake River (White Paper). Fort Collins: Colorado State University.

Pinsky, Malin L.; Jensen, Olaf P.; Ricard, Daniel; & Palumbi, Stephen R. (2011). Unexpected patterns of fisheries collapse in the world's oceans. PNAS Early Edition. Retrieved May 14, 2011 from http://www.pnas.org/cgi/doi/10.1073/pnas.1015313108

Rabanal, Herminio R. (1988). History of aquaculture. (ASEAN/SF/88/Tech. 7). ASEAN/UNDP/FAO Regional Small-Scale Coastal Fisheries Development Project. Retrieved May 14, 2011 from http://www.fao.org/docrep/field/009/ag158e/AG158E00.htm#TOC

Royce, William F. (1987). *Fishery development*. New York: Academic Press.

Shrestha, Ram K.; Seidl, Andrew F.; & Moraes, Andre S. (2002). Value of recreational fishing in the Vrazilian Pantanal: A travel cost analysis using count data models. *Ecological Economics, 42*, 289–299.

Webster, D. G. (2011). The irony and the exclusivity of Atlantic bluefin tuna management. *Marine Policy, 35*(2), 249–251.

Welcomme, Robin L., et al. (2010, September 27). Inland capture fisheries. *Philosophical Transactions of the Royal Society, 365*, 2881–2896.

Worm, Boris, et al. (2006, November 3). Impacts of biodiversity loss on ocean ecosystem services. *Science, 314*, 787–790.

Worm, Boris, et al. (2009, July 31). Rebuilding global fisheries. *Science, 325*, 578–585.

Food Webs

Food webs are a way to understand feeding interdependencies among species within an ecosystem. The degree to which such interdependencies remain intact determines the sustainability of a system. Humans are closely linked with natural systems and have therefore shaped food webs throughout the world. The conservation of food webs is essential to ensure sustainable resource use and to preserve ecosystem health.

A food web is a way to think about how plants and animals in an ecosystem are connected. Species are organized into food webs according to the resources they consume. Primary producers (plants) draw on inorganic nutrients from the soil and carbon dioxide from the atmosphere, primary consumers (herbivores) consume plants, secondary consumers (carnivores) consume herbivores, and decomposers (for example, bacteria) consume and recycle all other uneaten, dead organic matter.

A food web is depicted as a network of nodes interconnected by directional links. (See figure 1.) Nodes represent variables, such as species or nutrients, and directional links represent the feeding dependencies between consumers (i.e., the eaters) and resource species (i.e., the eaten). Energy is cycled through the system as it is repeatedly consumed and converted into edible tissue at each level of the chain and finally returned to the soil through decomposition. Most of the energy captured in food ends up lost as heat and indigestible waste during metabolic processes, and consumers convert an average of only 10 percent of chemical energy into tissue. This ecological inefficiency means that most of the biomass, or amount of living matter, in food webs tends to be tied up in plants and progressively less in herbivores and carnivores, creating a pyramid-shaped structure of food webs where plants are abundant and carnivores are rare. Abiotic factors, such as climate and nutrient availability, as well as biotic factors such as predation and competition, caused by living organisms, affect species' populations. As their environments change, species may adapt, for example, by changing their diet. The structure of a food web is quite flexible, as species turn to different food resources in response to environmental change.

Drivers of Ecosystem Stability

Prior to the mid-twentieth century, abiotic (nonbiological) factors were considered the dominant driver of ecological processes and species interactions. Weather and nutrient availability were considered to be the primary limitation on plant productivity, which in turn limited animal productivity. This so-called *bottom-up control* perspective was challenged in 1960, when zoology professors Nelson Hairston, Frederick Smith, and Lawrence Slobdkin proposed the green world hypothesis, which suggests that carnivores play a very important role in structuring ecosystems by consuming herbivores and thus limiting grazing on plants. So-called *top-down control*, in which predators limit herbivores and indirectly benefit plants, transformed the way that ecologists understand biological processes governing ecosystem functioning. Under top-down control, a change in a top predator species' abundance has a ripple effect on animals and plants lower down in the food web. This is known as a *trophic cascade*. Top-down, rather than bottom-up, control is now recognized as the stronger driver of food web processes in many systems, highlighting the important role that top carnivores play in maintaining food webs. The widespread extermination of large carnivores, such as wolves, bears, and cougars in North America, and the resulting shifts in vegetation due to overgrazing by herbivores (Beschta and Ripple 2009) is but one example of

Figure 1. An Example of a Food Web

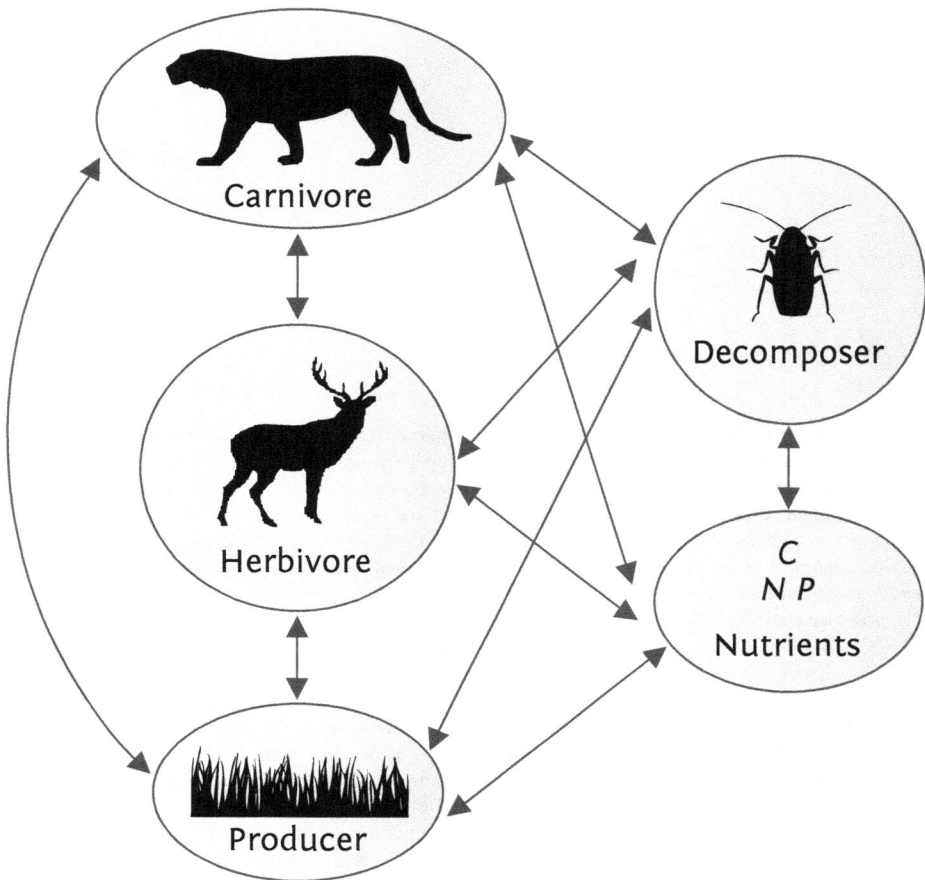

Source: Jennie R. B. Miller.

A classic food web showing trophic (nutritional) dependencies between a carnivore (tiger), herbivore (deer), producer (grass), decomposer (beetle), and nutrients (carbon, nitrogen, and phosphorus). The nodes are the species or the nutrients. The arrows depict interactions between species or nutrients and indicate lines of feeding dependency.

the large-scale effects on ecosystem structure arising from the dismantling of food webs.

The sustainability of an ecosystem is determined by both the species composition of its food web and the degree to which species are connected via feeding dependencies. While the sheer diversity of species alone was originally considered to be a fundamental driver of stability, it is now clear that the robustness of a food web is more related to its connectance, or the number and strength of feeding interactions between species, than to species diversity. In other words, the loss of any single species threatens the sustainability of an ecosystem because it reduces connectance between members. Not all

species affect stability equally, however. Food webs are composed of many weak and a few strong interactors. Intuitively, the strong interactors should play the greatest role in determining food web structure and ecosystem functioning and accordingly the loss of a strongly interacting species can drastically alter ecosystem function and threaten sustainability. For example, keystone predators are species that play significant roles in maintaining species diversity, and their removal from food webs can substantially shift interactions between other species, leading to changes in species composition, even to the point of ecosystem collapse. These species tend to be large, rare vertebrate consumers that are also most vulnerable to human threats from hunting and habitat degradation (Duffy 2003; Estes et al. 2011). Human-driven extinctions have dramatically skewed the structure of food webs by removing the higher-level species from them. These losses have caused complex rippling effects on ecosystem function, and as a result conservationists have made efforts to protect and restore these species; the reintroduction of wolves to Yellowstone National Park in the United States is one example. Counterintuitively, however, it is becoming apparent that ecosystems may be most sustainable when many weakly interacting species counterbalance the effects of a few strongly interacting species (McCann, Hastings, and Huxel 1998). In combination, many weakly interacting species can serve as important threads that hold ecosystems together. Efforts to restore and protect only the rare, strongly interacting species thus may be insufficient to protect the long-term sustainability of ecosystems.

Human Effects

People often recognize the importance of top predators in ecosystem sustainability only after the predator population is severely reduced or completely lost. For example, in the Scotian Shelf ecosystem off Nova Scotia, Canada, commercial overexploitation caused cod and other bottom-dwelling fish predator populations to crash in the late 1980s (Strong and Frank 2010). The decline in top fish predators generated a series of trophic cascade effects, including an increase in the abundance of prey species such as the northern snow crab and northern shrimp and a decrease in zooplankton due to the larger number of fishes. Top predator levels were also affected; large shark populations such as hammerhead sharks declined up to 99 percent in the western Atlantic due to loss of a significant food species. Many marine systems have recovered from overfishing disturbances within twenty years (Jones and Schmitz 2009), but despite a ban on cod fishing since 1993, the Nova Scotia ecosystem shows no sign of return to its original state. Unsustainable overfishing may have changed the interdependencies among species to the point that food web structure and dynamics shifted permanently, with unforeseen impacts on ecosystem function and species survival.

Besides direct impacts such as hunting and overfishing, humans have affected food web sustainability indirectly through urbanization. A prime example of this is so-called *mesopredator release* in southern California, where land fragmentation has caused a decline of the apex (top level) carnivore—the coyote—and the rise of smaller carnivore species (mesopredators). Apex predators can suppress mesopredators by killing them for food and by instilling fear, which changes mesopredator behavior, habitat use, distribution, and abundance. As a result of coyote decline, increasing numbers of mesopredators such as skunks, raccoons, foxes, opossums, and domestic cats killed more native birds, driving some species to extinction and significantly reducing bird diversity (Crooks and Soulé 1999). Even without directly removing top predators, human influences on the environment can change the nature and strength of feeding dependencies between species, generating unsustainable predation pressure and indirectly causing local extinctions.

Sustaining Human and Natural Systems

Conserving top predators is essential for maintaining ecosystem function and biodiversity, yet it can be logistically challenging and ethically controversial. Large carnivores range across broad, heterogeneous landscapes and seascapes that are often difficult to protect alongside expanding human development. Furthermore, top carnivores attack humans and livestock and generate substantial losses to the livelihoods of people living adjacent to protected areas (Woodroffe, Thirgood, and Rabinowitz 2005). They are viewed as pests in many parts of the world and are exterminated by the millions to protect game animals and domestic livestock. Understandably, farmers and villagers often resist conservation efforts involving the reintroduction of large carnivores. The continuing controversy regarding protection of wolves from renewed hunting now that they have rebounded in number in the greater Yellowstone Park area and adjacent states is a prime example of the social and political complications involved in conserving top predators in human-dominated landscapes (Musiani and Paquet 2004; Taylor 2011).

The preservation of food webs is crucial not only for ethical reasons but also for sustaining natural economies and preventing widespread poverty and food insecurity. The interconnectedness of large food webs can lead to unexpected declines in resources and shifts in livelihoods across the world. The collapse of the cod fishing industry in Nova Scotia, for example, led European Union fleets to shift to the oceans of western Africa, where they competed with local fisheries. The resulting decline in fish harvests by local fisheries motivated people in Ghana to turn increasingly to an unsustainable bushmeat (jungle animal) trade for income and food, which in turn caused sharp declines in forty-one tropical wildlife species (Brashares et al. 2004). Globalization has tightened the interdependency between international biological conservation, food security, and economic development to the extent that protection of food web linkages is now mandatory for the survival of human and natural systems alike.

Future Applications

As humans continue to influence natural processes, the food web is used increasingly as a tool for understanding how global ecosystems change. In addition to overexploitation and habitat fragmentation, climate change is now acknowledged as a major factor in trophic interactions. Climate shifts often create situations in which animals' life cycles are out of phase with their accessibility to food. For example, empirical evidence indicates that faster spring warming in the High Arctic of Canada has led to a mismatch in timing between the arrival of migratory caribou herds and the growth of their plant foods, which results in greater calf mortality (Post and Forchhammer 2007). In a European forest ecosystem, where spring has been arriving earlier over the past twenty years, caterpillars and predatory songbirds have adapted to the twenty-day advance, but the top predators—sparrow hawks—have not adjusted to their prey's life cycle development. Consequently, the hawks are unable to provide enough food for their young during breeding season. Fewer offspring survive, and the number of hawks overall is declining (Both et al. 2009). Analyzing food webs reveals that human influences on a single species can cause rippling effects throughout the food web network that change the ways individual animals behave and how populations change.

As the ecological and conservation sciences broaden from a species-specific to ecosystem-based understanding of natural processes, food webs will continue to aid in conceptualizing connections between species. Global interconnections between human economies and natural biodiversity indicate that people are intricately linked with environmental systems. To maintain ecosystem function and biodiversity, the integrity of food web structure, particularly the presence of top predators that drive top-down control of system processes, must be preserved.

Jennie R. B. MILLER and Oswald J. SCHMITZ
Yale School of Forestry & Environmental Studies

See also in the *Berkshire Encyclopedia of Sustainability* Agroecology; Biodiversity; Community Ecology; Complexity Theory; Ecosystem Services; Global Climate Change; Habitat Fragmentation; Hunting; Keystone Species; Microbial Ecosystem Processes; Nutrient and Biogeochemical Cycling; Plant-Animal Interactions; Population Dynamics; Regime Shifts; Resilience; Wilderness Areas

FURTHER READING

Beschta, Robert L., & Ripple, William J. (2009). Large predators and trophic cascades in terrestrial ecosystems of the western United States. *Biological Conservation, 142*, 2401–2414.

Both, Christiaan; Van Asch, Margriet; Bijlsma, Rob G.; Van Den Burg, Arnold B.; & Visser, Marcel E. (2009). Climate change and unequal phenological changes across four trophic levels: Constraints or adaptations? *The Journal of Animal Ecology, 78*, 73–83.

Brashares, Justin S., et al. (2004). Bushmeat hunting, wildlife declines, and fish supply in west Africa. *Science, 306*, 1180–1183.

Crooks, Kevin R., & Soulé, Michael E. (1999). Mesopredator release and avifaunal extinctions in a fragmented system. *Nature, 400*, 563–566.

Duffy, J. Emmet. (2003). Biodiversity loss, trophic skew and ecosystem functioning. *Ecology Letters, 6*, 680–687.

Estes, James A., et al. (2011). Trophic downgrading of planet Earth. *Science, 333*, 301–306.

Hairston, Nelson G.; Smith, Frederick E.; & Slobodkin, Lawrence B. (1960). Community structure, population control, and competition. *The American Naturalist, 94*, 421–425.

Jones, Holly P., & Schmitz, Oswald J. (2009). Rapid recovery of damaged ecosystems. *PloS One 4*, e5653.

McCann, Kevin S.; Hastings, Alan; & Huxel, Gary R. (1998). Weak trophic interactions and the balance of nature. *Nature, 395*, 794–798.

Musiani, Marco, & Paquet, Paul C. (2004). The practices of wolf persecution, protection, and restoration in Canada and the United States. *BioScience, 54*, 50–60.

Post, Eric, & Forchhammer, Mads C. (2008). Climate change reduces reproductive success of an Arctic herbivore through trophic mismatch. *Philosophical Transactions of the Royal Society of London, Series B, Biological Sciences, 363*, 2369–2375.

Strong, Donald R., & Frank, Kenneth T. (2010). Human involvement in food webs. *Annual Review of Environment and Resources, 35*, 1–23.

Taylor, Phil. (2011, July 28). Wolves move from endangered to hunted in rural Mont. Retrieved November 14, 2011, from http://www.nytimes.com/gwire/2011/07/28/28greenwire-wolves-move-from-endangered-to-hunted-in-rural-92931.html?pagewanted=all

Woodroffe, Rosie; Thirgood, Simon; & Rabinowitz, Alan. (2005). *People and wildlife: Conflict or coexistence?* Cambridge, UK: Cambridge University Press.

Forests

Forests contribute to the Earth's habitable conditions through maintenance of oxygen and carbon levels, water filtration, and nutrient cycling, among many other functions. Yet human harvesting of forest products and clearing of forests for roads, agriculture, and human settlements threaten the ecological services that forests provide.

Forests, which cover about one-third of the land area of the planet, hold an ambivalent place in the human imagination. People are dependent on intact forests for diverse ecosystem services, yet we clear forests for economic gain. In stories and legends, forests are depicted both as welcoming promised lands, rich in all varieties of natural resources, and as wild and untamed homes to dangerous beasts and threatening calamities. Reflecting these polyvalent attitudes, the US philosopher Henry David Thoreau (1817–1862) wrote, "The West of which I speak is but another name for the Wild; and what I have been preparing to say is, that in Wildness is the preservation of the world. Every tree sends its fibers forth in search of the Wild. Cities import it at any price. Men plough and sail for it. From forest and wilderness come the tonics and barks that brace mankind" (Thoreau 1851 in Nash 1990, 38).

Benefits of Forests

People use forests as sources of raw materials for manufacturing processes or subsistence livelihoods. Forests are logged for timber and for wood pulp to make paper (more than 40 percent of all trees logged are turned into paper). Many human populations rely on forests for hunting, non-timber forest products (e.g., fruits, nuts, vegetables, spices, meats, oils, saps, dyes, rubber, medicines, and raw materials for traditional arts and crafts), and forage for livestock. In addition to their material benefits, forests often serve as sites of inspiration, rejuvenation, and epiphany.

While forests are home to uncounted wild species (only a small percentage of the world's tropical rain forest species have been identified), they also provide a diverse array of benefits to humans. Known as "the lungs of the world" because of their essential functions in producing oxygen and moderating the water and carbon cycles, forests provide a diverse array of ecosystem services, including water catchment and filtration, nutrient cycling, soil formation, biodiversity preservation, biomass generation and decomposition, and carbon sequestration that create and maintain habitable conditions on Earth.

With increasing concerns about climate change due to anthropogenic carbon emissions, the essential activity of forests as carbon sinks, which take up atmospheric carbon in their growing processes, has come to be seen as very important. Through the Clean Development Mechanism of the Kyoto Protocol (1997), developing countries are paid to plant and maintain forests in order to capture atmospheric carbon and offset the carbon emissions of the wealthier countries that have made commitments to reduce greenhouse gases.

Dangers of Forest Loss

Recognizing the importance of forests to human and planetary well-being, scientists have warned of the dangers of forest loss. In 1996, the World Resources Institute estimated that just over half (53.4 percent) of the original forest area (meaning the forest area on the Earth as it is estimated to have existed 8,000 years ago, assuming current climactic conditions) remained (WRI 1996). During the 1990s, the UN Food and Agriculture Organization reported a global loss of 0.22 percent of forest cover annually, with the most significant losses in the tropics. Forests are cleared for the economic value of their timber, to create agricultural fields, and to expand human settlements. In addition, selective harvesting of timber or wild species, cattle grazing, and water diversion can degrade forests, decreasing soil fertility and biodiversity and lowering their productivity. Selective harvesting of timber that removes trees at the same pace at which they regenerate may be sustainable, and may, in some cases, enhance species diversity within forests by creating a wider variety of ecological niches. Such harvesting must be carefully monitored, through either scientific forest management or traditional taboos, to ensure that the pace of removal does not exceed the pace of regeneration.

Decreased quality or outright loss of forests causes a loss of the ecological goods and services that forests provide. Forest destruction is also a spiritual loss, as the richness of the living world is decreased and humans lose the opportunity to commune with various wild species. Meaningful relationships with the natural world may be necessary for optimum human functioning, according to the biophilia hypothesis proposed by the biologist Edward O. Wilson and the ecologist Stephen R. Kellert. Human well-being depends on the wholeness and coherence of living systems. The human species evolved in conjunction with natural processes that shaped human facilities and cognition. Only relatively recently, in evolutionary time, have humans begun living indoors. Because of our long co-evolution with other species, human beings are intrinsically dependent upon close relations with nature and natural processes for individual and general well-being.

In cultures throughout the world, forests have been places of refuge, contemplation and solace, where people could escape the hurly-burly of daily life to contemplate transcendent truths or interact with deities, spirits, and supranormal forces. The loss of forests reduces these opportunities, as well as opportunities for the aesthetic and uplifting appreciation of nature. In traditional and indigenous cultures, healing practices, spirituality, and world-view are often intertwined in systems that view illness as a spiritual imbalance or disruption. Forest loss also affects the ability to engage in these spiritual and traditional healing practices, as the necessary herbs for such practices are often only to be found in forests.

Scientists and policymakers in the developed countries have primarily been concerned with the loss of biological diversity, the loss of undeveloped wilderness, and the loss of recreational opportunities, while activists in the developing countries have been more concerned with the maintenance of rural livelihoods that depend on the extraction of timber and non-timber forest products for their subsistence, as well as preventing neocolonialist bio-prospecting and bio-piracy that would unfairly extract resources from the forests.

North America, South America, and Russia are home to the largest swaths of forests. Tropical rain forests, found in warm, low-elevation areas such as the Amazonian basin

of South America, are richest in biological diversity. The soil is relatively nutrient-poor, making it difficult to grow crops on the bare soil if the forest is cut down. If clearing the tropical rain forest continues at current rates, these forests will be gone by the end of the twenty-first century. Temperate forests are found in eastern North America and northeastern Asia. Because these forests are extremely productive, turning out 1,800 tons of vegetation per acre (as compared with 185 tons per acre in a tropical forest), most of the original temperate forest has long since been cleared. At the highest latitudes and altitudes of North America and Eurasia, the alpine and boreal forests are found, and they represent the largest terrestrial biome. Precipitation occurs mainly in the form of snow. Boreal forests are extensively logged and may vanish unless logging is limited. The boreal forests, circling the subarctic zone around the Northern Hemisphere, are estimated to constitute 25 percent of the world's remaining forest. Thus, although forested areas continue to exist in many climate zones of the Earth, forest resources are being depleted for human use, threatening overall sustainability of life on the planet.

Elizabeth A. ALLISON
University of California, Berkeley

FURTHER READING

Eisenberg, Sheryl. (2004, February). Taking trees personally. *This Green Life*. Retrieved March 29, 2009, from http://www.nrdc.org/thisgreenlife/0402.asp

Food and Agriculture Organization of the United Nations. (2003). *State of the world's forests, 2003*. Rome: United Nations. Retrieved March 29, 2009, from http://www.fao.org/DOCREP/005/Y7581E/Y7581E00.HTM

Kyoto Protocol: Mechanisms: Clean development mechanism. Retrieved May 22, 2009, from http://unfccc.int/kyoto_protocol/mechanisms/clean_development_mechanism/items/2718.php

National Aeronautics and Space Administration (NASA). (1998). Tropical deforestation. *The Earth Science Enterprise Series*: FS-1998-11-120-GSFC. Retrieved March 29, 2009, from http://www.iwokrama.org/library/pdfdownload/NASAdeforestation.pdf

Rosenzweig, M. L., & Daily, G. C. (2003, June 6). Win-win ecology: How the Earth's species can survive in the midst of human enterprise. *Science 300*(5625): 1508–1509.

Thoreau, Henry David. (1851). Walking. Reprinted in Roderick F. Nash, (Ed.) (1990), *American environmentalism: Readings in conservation history* (3rd ed.),p. 38. New York: McGraw Hill.

United Nations Framework Convention on Climate Change: Kyoto Protocol, Clean Development Mechanism. (Adopted 1997; entered into force 2005). Retrieved May 22, 2009, from http://cdm.unfccc.int/index.html

Wilson, Edward O., & Kellert, Stephen R. (Eds.). (1993). *The biophilia hypothesis*. Washington, DC: Island Press.

World Resources Institute. (2009). Earthtrends: Environmental information. Retrieved May 27, 2009, from http://earthtrends.wri.org/

World Resources Institute. (1996). Forests, grasslands and drylands—Forest extent: Forest area (current) as a percent of original forest area. Retrieved August 6, 2009, from http://earthtrends.wri.org/searchable_db/index.php?theme=9&variable_ID=313&action=select_countries

Glaciers

Glaciers are important because they regulate fresh-water supplies, affect sea level, and influence natural hazard vulnerability. They are also cultural icons, tourist destinations, and sites for nature conservation. Glaciers can also trigger outburst floods and ava-lanches. Most glaciers have shrunk considerably since the mid-nineteenth century. Glacier retreat caused by global climate change threatens societies worldwide because glacier runoff provides water for drinking, irrigation, and hydroelectricity.

Glaciers are masses of ice that flow downhill. They form when snow remains on the ground throughout summer so that, over time, the weight of successive snow layers compacts the snow and eventually turns it to ice. Ice becomes a glacier after it begins to move internally (deform) and slide downhill, which generally requires an ice thickness of at least 35 meters. Glaciers cover approximately 10 percent of the Earth's land surface, and glacial ice represents 80 percent of the world's freshwater. Antarctica contains 90 percent of the world's glacial ice, and Greenland contains 9 percent, leaving just 1 percent in all the world's other glaciers. Of course, glacial cover-age changes through time: at the height of the last ice age twenty thousand years ago, glaciers covered as much as 30 percent of the Earth (Hambrey and Alean 2004).

Since the end of the Little Ice Age (a cooler period worldwide from approximately 1350 to 1850), the vast majority of glaciers worldwide have shrunk considerably, and some small glaciers have disappeared completely. The most notable glacier retreat has occurred in areas with these smaller glaciers, such as on Africa's Mount Kenya and Kilimanjaro, as well as in parts of Bolivia and Ecuador; glacier retreat in New Zealand and parts of Alaska, however, has not been as uniformly dramatic.

Concerns about global warming have led many to worry about future glacier loss, which would affect freshwater supplies, sea level, natural hazard vulnerability, nature conservation, cultural icons, tourist destinations, and climbing terrain.

Glacial Ice as Natural Resources

Glaciers offer many benefits to people, including practical or economic resources. Glacial ice has historically been used to cool drinks and refrigerate foods. The use of mountain snow and glacial ice dates back thousands of years in China, the Middle East, North Africa, and Europe. By the sixteenth century Europeans used ice consistently, and major markets developed in London, Paris, Granada, and Florence. In the 1600s Spanish colonizers in the Andes mountains of South America imposed an ice *mita*, a system of forced labor that sent indigenous people to collect glacial ice for wealthy Spaniards living on South America's coast. In Europe, Norway was a leading exporter of glacial ice before the late 1800s, when mechanical refrigeration took over. Some Japanese still use Alaskan glacial ice to cool their drinks. Peruvians continue to eat flavored snow cone–like *raspadillas* made from glacial ice transported by mule or on the backs of local ice collectors 2,000 vertical meters down mountains to towns.

Glaciers as Water Resources

People rely on glacial runoff for the water to drink, to irrigate, and to generate hydroelectric energy. Glaciers are ideal for water storage. Ice (and therefore water) accumulates during cold, wet winters. Glaciers then release water when summer temperatures rise and during dry seasons when little precipitation falls. Glaciers thus regulate annual water flows and provide water when people need it most. Shrinking glaciers threaten natural resource sustainability primarily by disrupting the distribution of water supplies over the course of a year. There is still scientific debate and uncertainty about the precise impact of glacier recession on water supplies, as well as on the timing of hydrologic impacts from shrinking glaciers based on future climate change scenarios. This uncertainty is particularly high for Himalayan glacier-hydrology dynamics, and especially for populations in China and the hundreds of millions of inhabitants using water from the Ganges, Yangzi (Chang), and Huang (Yellow) rivers that may not be as significantly affected by glacier retreat as previously thought. The Indus River, however, that flows from China to India and Pakistan will be markedly diminished by glacier retreat. Loss of glaciers is also expected to cause significant dry-season declines in water supplies in the Andes, where residents of Lima, La Paz, and Quito depend on the water from glaciers, and a significant portion of Andean energy comes from hydroelectricity. In western North America, there has also been a decline in late-summer stream flow in glacier-fed watersheds. As these water supplies and annual hydrologic distributions fluctuate in the future, social struggles and water conflicts will almost undoubtedly occur as groups jockey to defend water rights in glacier-fed watersheds.

Glaciers and People

Glaciers have shaped cultures and guided people's spirituality. For at least two thousand years, peoples of South America have venerated the life-giving glaciated peaks of the Andes. Today religious pilgrims annually climb to the glaciers of Mount Ausangate in Peru, where they collect sacred, healing, and purifying glacial ice. Glaciers have also played a prominent role in the culture of Tlingit and Athapaskan peoples in northwestern Canada. Oral histories reveal that they viewed glaciers as sentient beings that affected hunting patterns and trade routes, as well as influencing social behavior, language development, place naming, and spirituality.

Glaciers also lure recreationists. Since the late 1700s mountaineers have climbed glaciers to reach the world's highest peaks. The link between glacier study

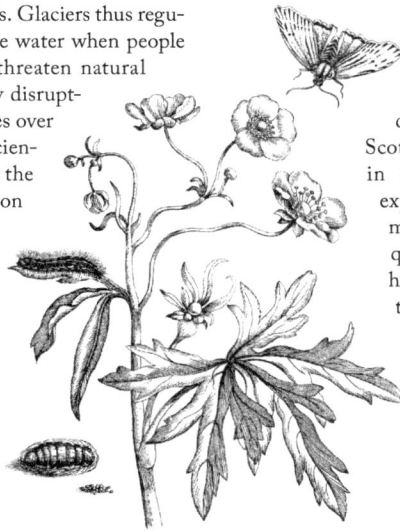

(glaciology) and mountaineering has deep roots: early mountaineers carried scientific instruments and made geological investigations during their climbs. For example, the Scottish physicist James Forbes—who in 1843 made the first attempts to explain physical dynamics of glacial ice motion—found that studying glaciers quenched his intellectual thirst and hiking on glaciers fulfilled his emotional passions. The appeal of glaciers has also driven people to the Earth's poles. Curious explorers have visited the Arctic since the Greek sailor Pytheas attempted a voyage there twenty-five hundred years ago. The first landing in Antarctica was in 1895, and the first visit to the South Pole (by Norwegians) was in 1911. Recreationists' interest in Antarctica increased after 1958, when New Zealander Sir Edmund Hillary made the first traverse of Antarctica by land.

Protecting and Sustaining Glaciers

National parks and other nature preserves often concentrate on protecting glaciated environments for recreation, tourism, and species conservation. Glacier National Park in the United States, Ilulissat Icefjord in Greenland (Denmark), Sagarmatha (Mount Everest) in Nepal, Mount Kilimanjaro in Tanzania, Mount Huascarán in Peru, Los Glaciares National Park in Argentina, and Westland Tai Poutini National Park in New Zealand are just some of the world's protected areas that were created to protect glaciated landscapes. Many of these have also become World Heritage Sites due to the increasingly recognized importance of glaciers worldwide. (World Heritage Sites are UN-designated places of cultural or physical significance.)

Since the late-twentieth century, some efforts have been made to prevent glaciers from shrinking as a consequence of global climate change. Proposals to cover Kilimanjaro with white tarps, for example, emerged in recent years, and ski resorts in Canada, Austria, and Switzerland have covered glaciated ski terrain with PVC foam and white fleece to prevent the ice from melting. In Ladakh, India, the so-called Glacier Man, Chewang Norphel, has built ten artificial "glaciers" to help local farmers. By pooling annual snowmelt and rainfall, Norphel was able to create bodies of ice that acted like glaciers by releasing water in late spring and early summer when farmers needed it to grow wheat, barley, and peas. In 2010, the World Bank funded a much-criticized project to paint Andean mountaintops white to change the albedo so the mountains would reflect sunlight instead of absorbing its heat. The painted landscape would thus, theoretically, prevent adjacent glaciers from melting. Schemes to save glaciers worldwide demonstrate the diverse and important roles glaciers play for farming, tourism, and cultural meaning.

Glacial Hazards

Glaciers can also pose serious threats to people. During the Little Ice Age (c. 1350–1850), global temperatures fell by 1–2°C. In response, glaciers generally advanced worldwide, occasionally creating problems for nearby inhabitants when glaciers overtook their pastures, fields, irrigation systems, bridges, homes, and even entire villages. In 1663, Alps residents below the Aletsch glaciers pleaded with nearby Jesuit priests to perform ceremonies that would turn back the wicked glaciers. When the Black Rapids glacier in Alaska surged in 1936–1937, a wall of ice began racing at 66 meters per day toward local residents; today a surge in this glacier could break the Alaskan pipeline.

Catastrophes also occur when glaciers generate outburst floods. Advancing glaciers from side valleys often dammed main valleys in the Alps during the Little Ice Age. In 1595 the Giétroz glacier dammed the Val de Bagnes in Switzerland; when the ice dam burst, it killed five hundred people. In Argentina's Mendoza region advancing glaciers have produced ice-dammed lakes and catastrophic floods since 1788, and a 1934 flood killed many people, destroyed bridges, and wrecked 13 kilometers of railway. Retreating glaciers can also trigger outburst floods. A deadly example occurred in Huaraz, Peru, in 1941, when a melting glacier formed a lake that burst through its moraine dam and killed five thousand people. Efforts were made to drain these glacial lakes and prevent catastrophic flooding in the Alps during the Little Ice Age, in the Andes from the 1940s to the present, and in the Himalayas since the 1980s.

Ice avalanches pose additional threats. In 1970 fifteen thousand people perished when an earthquake triggered an avalanche in Peru's Cordillera Blanca. When glacier-covered volcanoes erupt, glacial ice melts quickly and creates mudflows called "lahars" (avalanches consisting of ice, mud, water, and other debris). One of the worst cases occurred in 1986 when Colombia's Nevado del Ruiz erupted and produced a lahar that killed thirty thousand residents.

Glaciers and Climate Change

Because temperature and precipitation are the main factors that determine their size, glaciers make excellent climatic indicators. Glacier advances in Europe around 1600–1610, 1690–1700, the 1770s, 1820, and 1850 reveal the coolest periods of the Little Ice Age, whereas glacier retreats during the Medieval Warm Period (c. 800–1200) and during much of the twentieth century demonstrate warmer periods. To understand climate history, glaciologists have drilled ice cores in glaciers, which store data about precipitation, temperature, atmospheric conditions, volcanic eruptions, and winds. Scientists have drilled cores worldwide—from Kilimanjaro in Tanzania to Huascarán in Peru to Everest on the Tibetan Plateau to Vostok in Antarctica. Ice cores from Greenland, where drilling began in 1958, offer accurate climate records going back 110,000 years. In Antarctica, a 2004 collaborative European project at Dome C extracted an ice core with 740,000 years of climate history. These records have become particularly important in recent discussions of global warming because the glaciers storing climatic history may also hold the key to understanding global climate in the future.

Mark CAREY
University of Oregon

See also in the *Berkshire Encyclopedia of Sustainability* Aquifers; Greenhouse Gases; Mountains; Oceans and Seas; Rivers; Water Energy; Water (Overview)

FURTHER READING

Alley, Richard. (2000). *The two-mile time machine: Ice cores, abrupt climate change, and our future.* Princeton, NJ: Princeton University Press.

Barry, Roger G. (2006). The status of research on glaciers and global glacier recession: A review. *Progress in Physical Geography, 30*(3), 285–306.

Carey, Mark. (2010). *In the shadow of melting glaciers: Climate change and Andean society.* New York: Oxford University Press.

Carey, Mark. (2007). The history of ice: How glaciers became an endangered species. *Environmental History, 12*(3), 497–527.

Cruikshank, Julie. (2005). *Do glaciers listen?: Local knowledge, colonial encounters, and social imagination.* Vancouver: University of British Columbia Press.

David, Elizabeth. (1994). *Harvest of the cold months: The social history of ice and ices.* London: Michael Joseph.

Fagan, Brian. (2000). *The Little Ice Age: How climate made history, 1300–1850.* New York: Basic Books.

Grove, Jean. M. (1988). *The Little Ice Age.* New York: Routledge.

Hambrey, Michael, & Alean, Jürg. (2004). *Glaciers.* New York: Cambridge University Press.

Immerzeel, Walter W.; van Beek, Ludovicus P. H.; & Bierkens, Marc F. P. (2010). Climate change will affect the Asian water towers. *Science, 328*(5984), 1382–1385.

Le Roy Ladurie, E. (1971). *Times of feast, times of famine: A history of climate change since the year 1000.* New York: Doubleday.

Mulvaney, K. (2001). *At the ends of the Earth: A history of polar regions.* Washington, DC: Island Press / Shearwater Books.

Orlove, Ben; Wiegandt, Ellen; & Luckman, Brian. (Eds.) (2008). *Darkening peaks: Glacier retreat, science, and society.* Berkeley: University of California Press.

Global Climate Change

The factors and processes involved in climate change are many and complex, ranging from fluctuations in the Earth's orbit to changes in biota. The Earth's climate has always been in flux, but indications are that human impacts from industrialization, land-use change, and growing population are speeding a warming of the planet that could have substantial effects on ecosystems and the services they provide.

Global climate change refers to alteration in climate (temperature, precipitation, and wind patterns) over a significant area lasting for an extended period. The complex set of processes involved in climate change includes impacts from land use (ice coverage and vegetation shifts, deforestation, development, urbanization, infrastructure deployment), natural and human-induced forcing factors, and feedback processes within the climate and Earth systems. It has long been recognized that the Earth's climate is in constant flux and that human activity can induce change, but the apparent complexity and underlying drivers of climate change have only come to light during the past century aided by technological advances and accumulated evidence. Population and economic growth are the major anthropogenic (human-generated) drivers of change in natural resources, land use, and their climate feedback processes. This discussion addresses forcing factors for global climate change and associated feedback mechanisms.

Climate

Climate is primarily regulated by the amount of energy absorbed and dissipated by the Earth's surface. Incoming shortwave radiation emitted by the sun passes through the atmosphere and strikes the Earth's surface. There it is either absorbed or is reflected as longwave radiation, depending on the albedo (the reflective property of the surface, including cloud cover) at that location. Some of the reflected radiation is trapped by greenhouse gases in the atmosphere (e.g., carbon dioxide [CO_2], methane [CH_4], nitrogen oxides [NOx], and water vapor), resulting in what is known as the greenhouse effect. This effect is largely responsible for Earth's average surface temperature of approximately 15°C, to which we have grown accustomed; removal of greenhouse gases would reduce the average surface temperature to about −18°C.

The amount of energy Earth receives from the sun varies with latitude. The sun's rays hit the equator directly, causing tropical regions to receive a large amount of energy. At higher latitudes, the same incoming solar radiation is distributed over a larger surface area of the Earth, creating the temperate and polar zones. The uneven distribution of heat across land and oceans fuels atmospheric circulation (Hadley circulation), thereby creating climate and precipitation patterns across the planet. This translates into weather patterns that develop in the lower atmosphere and are driven by incoming heat from the sun, the Earth's rotation, and heat stored in oceans and the atmosphere. The storage capacity of heat in the atmosphere is a function of the relative amount of the incoming radiation that can be absorbed by the different greenhouse gases in the atmosphere. Large water masses (oceans) have a high capacity to store heat and therefore cool and warm very slowly.

The temperature differences between land, oceans, and air ultimately drive climate and explain temperature and precipitation patterns along the Earth's surface in a predictable fashion. As a consequence of this predictability, biota adapts to geographical locations of the Earth in highly recognizable forms such as tropical, temperate, arid, or polar biomes. In turn, these biological systems can also affect climate as they exchange large amounts of greenhouse gases with the atmosphere, particularly CO_2 and water vapor. Therefore, any factors that impact the biosphere–atmosphere energy balance may result in relatively rapid climate change.

Recent Change

Satellite, weather balloon, and ground observations all agree that there has been a steady warming trend in the Earth's surface temperatures and that it has been more apparent over the course of the nineteenth and twentieth centuries. Based on meteorological data, the twentieth century can be divided into three sections: early twentieth-century warming, a mid-century cooling episode, and late twentieth-century warming (Anderson, Goudie, and Parker 2007).

Descriptions of early twentieth-century warming documented the changing time periods marked by the occurrence and intensity of first and last snowfalls or ice covers: for example, they noted how the snow period declined from 150 days to 113 days in London, or when the period of time during which ice cover in the Arctic Sea prevented navigation shortened from 12–13 weeks to 3–4 weeks per year, or when polar ice thickness declined 20–40 percent depending on location. During the early twentieth century sea temperature records reveal about a 1°C–2°C increase until the 1960s in northern latitudes. These increases in temperature were corroborated by independent biological observations, including the northward expansion of cod, halibut, or haddock in Greenland, displacement of fish by warm-adapted species in the southern limits (though overfishing has contributed somewhat to these effects), and the northward range shifts of plant species and birds, including the invasion of tundra by trees between 1920 and 1940. This warming period also impacted agricultural and silvicultural practices, as the number of growing days increased, and cultivation of rye, barley, or oats expanded into high latitudes in Scandinavia (expansion not caused by breeding) (Anderson, Goudie, and Parker 2007).

The mid-century cooling period occurred between about 1945 and 1970 on land and 1955 and 1975 in the oceans. Unlike the early twentieth-century period, when 85 percent of the Earth's surface experienced warming, during the mid-century cooling period 80 percent of the total Earth surface area experienced cooling. During this period glaciers stopped retreating, snowbanks were formed in the Canadian Arctic, snowfall increased in Europe, Baltic Sea ice increased, and the plant-growing season was documented to be shortened in parts of northern Europe.

The late twentieth-century warming period is characterized by a rapid increase in temperatures over continents at mid latitudes (40–70 degrees north). Temperatures rose 0.6°C in about two decades, the fastest and largest increase in temperature known over the last thousand years. This warming trend has primarily affected night-time temperatures as increased cloud cover has contributed to reduced diurnal temperature oscillations. During this current period similar climatic and biological trends that characterized the early twentieth-century warming period have been observed. Glacier retreat and melting of the permafrost (at about 4–5 kilometers a year) have been particularly well documented. Also, the onset of spring for both plant and animal life is occurring five to eleven days earlier than indicated in the historical record.

In addition to increases in temperature during the twentieth century, global precipitation has increased by about 2 percent in response to the higher evaporation rates of ocean waters. The magnitude of rainfall events has increased in many areas of the Northern Hemisphere and Australia. The increase in precipitation at northern latitudes is contrasted with decreased precipitation and increased aridity at low latitudes, particularly in northern Africa and Asia, indicating that climate shifts will not be uniform. Much of the variability observed in precipitation patterns is also related to the El Niño Southern Oscillation (ENSO), the complex of warm ocean current and associated atmosphere that influences continental climate in many regions of the world.

Multiple lines of evidence indicate changes in climate over the last 150 years. Debate continues, however, on what is causing the temperature and precipitation changes since the late nineteenth century. Changes in atmospheric chemistry due to human activities that can lead to both warming (greenhouse gases) and cooling (aerosols) seem to explain a large part of the surface temperature oscillations at short-term scales.

Natural Drivers

The Earth's climate has continuously changed during the planet's history. In the past, climate was largely impacted by natural physical, chemical, and biological processes and the feedback between these. Tectonics, which creates land and moves continents on the Earth's surface, clearly influences climate, but because continental movement is very slow, tectonics alter climate over tens of millions of years. Over the last 2–3 million years climate has changed more rapidly, with spans of tens of thousands of years forming cold (glacial) periods and warm (interglacial) periods. These climate changes can largely be explained by planetary forcing agents that affect the amount of

incoming energy from the sun hitting the Earth's surface. The theory of orbital forcing developed by the
Scottish scientist James Croll in the 1860s and advanced
by the Serbian civil engineer and mathematician Milutin
Milankovitch in the 1920s describes how the eccentricity, axial tilt, and precession of the Earth's orbit in relation to the sun drive glacial–interglacial variations
(Imbrie and Imbrie 1979). Slight variations in these
parameters directly impact the amount of solar radiation
reaching the Earth and subsequently impact the seasonality and location of solar energy.

The Earth and all other planets in our solar system
orbit the sun in an elliptical manner. The eccentricity of
the orbit, or the departure of the ellipse from circularity,
is determined by the interactions between the gravita-
tional fields of Jupiter and Saturn. The ellipticity of
Earth's orbit varies from 0 to 5 percent on a cycle of
roughly 100,000 years. Variations in eccentricity account
for how far the Earth is from the sun and have contributed to historic glacial regimes. The angle of Earth's axial
tilt in relation to its plane of orbit around the sun is
responsible for seasonal variation in daylight and temperature. Currently the axial tilt of the Earth is close to
23.5 degrees and decreasing; Earth's tilt naturally varies
from approximately 21.4 degrees to 24.5 degrees on a
roughly 41,000-year cycle. Additionally, the Earth's precession governs how the Earth wobbles as it spins on its
axis and operates on a periodicity of about 23,000 years,
further modulating seasonality. Evidence from deep-sea
sediments and ice cores suggest considerable climate variability is associated with orbital forcing (Imbrie et al. 1992).

Regarding shorter time scales, it has been hypothe-
sized that shifts in the quality (via changes in ultraviolet
[UV] range) and quantity (via sunspots) of solar radiation
at the Earth's surface can also result in changes in climate (Lean 2010). Research suggests that the number of
sunspots varies on a roughly eleven-year cycle and can
alter solar output by approximately 0.01 percent. During
periods of high sunspot activity, the Earth receives
increased radiation compared to periods with low activity. It is thought that since 1750 increased solar irradiance
has been responsible for a positive radiative forcing of
0.06 to 0.30 watts per square meter (W/m^2) (IPCC
2007). This is sufficient to contribute to moderate
increases in temperature in the upper atmosphere but
cannot account for most of the observed increases in surface temperatures.

Volcanic eruptions may also play an important role in
the Earth's climate through two primary pathways: first
through the emissions of CO_2 and other greenhouse gases
into the atmosphere and second by emissions of aerosols
(suspensions of fine particles in gas) such as ash and sulfur
gases. Water vapor and CO_2 are the primary greenhouse
gases emitted and represent between 50–90 percent and

1–40 percent of annual volcanic emissions, respectively.
The water vapor dissipates from the atmosphere rapidly,
resulting in a negligible effect on climate, while the magnitude of CO_2 from volcanic origins is less than 1 percent
of annual CO_2 emissions (Gerlach 1991). Additionally, ash
and sulfur gases are projected into the stratosphere and can
contribute to global cooling. These aerosols reflect incoming radiation back to space, leading to cooling of ground
surface temperature. Volcanic ash is usually removed rapidly (within one month after the eruption) from the atmosphere by sedimentation (Pinto, Turco, and Toon 1989).
Sulfur gases from volcanic activity represent about 36 percent of the annual tropospheric sulfur emissions (Graf,
Feichter, and Langmann 1997) and are largely responsible
for the climatic effects associated with eruptions because
of their longer residence times in the atmosphere and their
role of scattering solar radiation back to space.

Additionally, natural fluctuations in Earth's albedo
resulting from shifts in land or cloud cover can impact
climate patterns by altering the amount of solar radiation
that is reflected or is absorbed by the Earth's surface. For
instance, increased snow cover can increase reflectance
and thereby alter the Earth's albedo, resulting in further
cooling. In contrast, increased vegetation can result in
what is called vegetative forcing, which lowers the land
surface albedo and results in increased absorption of heat,
thereby raising surface temperatures.

As previously mentioned, the Earth's atmospheric and
oceanic conditions are closely coupled, and thus alterations in patterns of oceanic circulation can have considerable impact on global climate. The combined effects of
heating/cooling and salinity drive the oceanic currents to
circulate water throughout the Earth's oceans. This is
known as thermohaline circulation and is responsible, for
instance, for warming the North Atlantic regions by as
much as 5°C. Evidence suggests that thermohaline circulation has been disrupted a number of times in the
past, resulting in considerable alterations of regional temperatures. For example, evidence suggests the Younger
Dryas, a millennium-long cold period about twelve thousand years ago, at the beginning of the Holocene, may
have been triggered by the release of freshwater into the
North Atlantic, altering ocean circulation (Broecker
1997). A shift in the ocean's thermohaline circulation
could occur with increased precipitation at higher latitudes, which would reduce salinity and thereby disrupt
circulation (Stocker and Schmittner 1997).

Anthropogenic Drivers

Humans, like most organisms, modify their environmental surroundings. As such, it is logical that the magnitude
of modification by humans has grown in conjunction

with population. The rapid growth of the human population has been fueled by the consumption of natural resources. Extraction of these natural resources is both energy and land intensive. Recently an increasing body of scientific literature suggests that there is compelling evidence that human activities are modifying forcing factors that influence climate (IPCC 2007). These human impacts are due in large part to the increased emission of greenhouse gases through the burning of fossil fuels (such as coal and petroleum), industrial activity, land-use change, and deforestation practices, all of which became prevalent during the industrial development of the past 250 years.

Human activities in large part bear responsibility for the increases in atmospheric greenhouse gases, which alter the Earth's energy budget through a process known as radiative forcing. Increased concentrations of these gases in the atmosphere contribute to global warming by absorbing energy reflected from Earth and re-emitting this energy, resulting in a net increase of energy. Humans have increased atmospheric CO_2 concentrations through fossil fuel combustion (estimated at 7.2 gigatons of carbon [GtC] per year from 2000 to 2005) and to a lesser extent by land clearing (estimated at 1.6 GtC per year during the 1990s) (IPCC 2007). These emissions have increased global CO_2 concentrations from preindustrial-era levels of about 280 parts per million (ppm) to about 389 ppm in 2011, far exceeding the range (180 to 300 ppm) from the last 420,000 years as determined from ice cores (Petit et al. 1999; IPCC 2007). Evidence suggests the increased atmospheric CO_2 has contributed to the global temperature increase from 1850/1899 to 2005 of an average 0.76°C (range of 0.57°C–0.95°C) (IPCC 2007). These patterns are particularly concerning due to the fact that CO_2 has a one-hundred- to two-hundred-year residency in the atmosphere, resulting in potentially long-lasting consequences. Three lines of evidence show that CO_2 increases are anthropogenic (Prentice et al. 2001). First, atmospheric oxygen is declining in line with CO_2 combustion. Second, the isotopic signature of fossil fuel (lack of carbon 14 [^{14}C] and depleted carbon 13 [^{13}C]) is detected in atmospheric measurements. Finally the increase in CO_2 is more rapid in the Northern Hemisphere where the majority of fossil fuels are combusted.

Humans have also contributed to the increase in a variety of other trace gases (principally methane, nitrous oxide, and halocarbons) that may have radiative forcing effects that are comparable to and higher than that of CO_2. Humans are now responsible for nearly 70 percent of annual methane atmospheric accumulation as a result of agricultural practices (i.e., livestock farming and rice cultivation), fossil fuel combustion, and decomposition associated with landfills. This has led to an increase in global methane concentrations from about 320–715 parts per billion (ppb) during the preindustrial period to 1,774 ppb in 2005 (IPCC 2007). Although methane has a relatively short residence time in the atmosphere (about twelve years) compared to CO_2, it exhibits 3.7 times more global warming potential per mole (Lashof and Ahuja 1990). Nitrous oxides have also increased from preindustrial levels of about 270 ppb to 319 ppb in 2005, primarily due to the burning of fuels at high temperatures such as in factories and cars (IPCC 2007). Finally, concentrations of halocarbons have increased significantly due to their use in synthetic organic compounds. The combined impacts of these trace gases have been estimated between 0.88 and 1.08 W/m², which constitutes nearly 60 percent of the radiative forcing of CO_2 (IPCC 2007).

Unlike CO_2 and other greenhouse gases that warm the atmosphere via positive radiative forcing, aerosols cool the atmosphere by reflecting incoming radiation (as in the case of large volcanic eruptions). Aerosols can contain a broad collection of particles with different chemical properties causing each to interact uniquely with the atmosphere. They can attract water and serve as cloud condensation nuclei, resulting in more diffuse clouds, which reflect more solar radiation. Sulfur dioxide produced from fossil fuel combustion and the burning of vegetation is the primary atmospheric aerosol. Aerosols are not long-lived in the atmosphere and are generally localized to the region of production. Although anthropogenic aerosol emissions have declined in North America and Europe due to more stringent regulations, emissions have increased in Asia as urbanization has increased.

Human population growth has relied on the widespread transformation of the Earth's surface to provide necessary resources, and current consensus suggests that humans have transformed or degraded somewhere between 39 and 50 percent of the Earth's surface (Vitousek et al. 1986 and 1997). Land surface change through deforestation, reforestation, and urbanization alters the albedo of the Earth's surface, impacting the amount of energy absorbed. Estimates indicate that the impacts of land transformation on Earth's albedo accounts for a loss of 0.4 W/m² (IPCC 2007), and therefore affecting the energy balance of the Earth's surface.

Feedback Mechanisms

To further complicate the variable climate system on Earth, global conditions are also modified by natural and human-induced feedback mechanisms, which operate on complex spatial and temporal scales. Climate feedback can be either positive or negative: positive feedback processes magnify an effect, causing increased warming or cooling, while negative feedback processes dampen change in climate.

An important feedback mechanism impacting climate is the flux of CO_2 into and from oceans; when global temperatures become warmer, CO_2 may be released from oceans. Increasing CO_2 concentrations may amplify warming by enhancing the greenhouse effect. When temperatures become cooler, CO_2 enters the ocean and contributes to additional cooling. During the last 650,000 years, CO_2 levels have tended to track glacial cycles: during warm interglacial periods, CO_2 levels have been high, and during cool glacial periods, CO_2 levels have been low.

Another important positive feedback process is the natural emission of CO_2 from soils (soil respiration), specifically in boreal ecosystems. As boreal ecosystems experience warming, they will release the large stocks of carbon that are currently immobilized (sequestered) in soils by frost, leading to further increases of CO_2 in the atmosphere. Furthermore, soil respiration has been shown to be correlated with temperature and moisture, such that increases in soil temperature in conjunction with moisture may result in increased natural CO_2 emissions rates from ecosystems (Wildung, Garland, and Buschbom 1975).

Alterations to the Earth's surface can also result in complex feedback effects on climate. Ice-albedo feedback refers to the lower albedo that snow and ice have compared to ground and vegetation, resulting in increased reflectance of energy into space. Periods of low temperatures allow snow cover to last for a longer duration, causing increased reflectance and the cooling of Earth's climate, which in turn can result in further expansion of ice cover (positive feedback). This process can also work in reverse, whereby reduced ice coverage creates feedback in which the Earth's surface warms, resulting in glacial recession. Current scientific consensus indicates that glaciers and ice caps have been losing mass, particularly since the early 1960s (Kaser et al. 2006).

Water vapor feedback in conjunction with other processes can amplify climate change. Despite its short residence time in the atmosphere, water vapor is a potent greenhouse gas. It has a heat-amplifying effect and tends to increase in conjunction with temperature, thus creating positive feedback. But water vapor also forms clouds that block incoming radiation, resulting in the cloud negative feedback effect (Ramanthan et al. 1989). A negative feedback effect by clouds is thought to be partly responsible for the observed moderate increases in surface temperature, which are less than expected given the accumulation of greenhouse gases in the atmosphere.

A study investigating feedback between climate and boreal forest vegetation cover during the Holocene epoch highlights the significance of orbital forcing, vegetation shifts, and feedback between these two parameters (Foley et al. 1994). The study suggests that orbital forcing (i.e., shifts in Earth's eccentricity, tilt, and precession), while capable of increasing global temperatures by 2°C during the mid Holocene, was not solely responsible for the observed warmer temperatures during the period. Instead, orbital forcing paired with its positive feedback effect on the northward expansion of boreal forests in high latitudes was likely to have contributed to the warmer temperatures observed. While not all feedback mechanisms are understood, research indicates that they are important in determining Earth's climate.

Sustainability, Biodiversity, and Resource Management

It is likely that human impacts on the global climate will continue, as global population and per capita energy consumption continue to rise. These impacts may be beyond the capacity for many ecosystems to adapt naturally, leading to losses of biodiversity and impacting ecosystem function and services. Consequently, climate change will be a major driver of decision making in natural resource management.

Global climate change can impact ecosystems and associated organisms in many ways. For instance, plants and animals may shift their ranges dramatically, moving as much as 6.1 kilometers per year toward the poles (Parmesan and Yohe 2003). The ability of individual species to respond to climate change will, however, likely be impeded by human-induced land habitat fragmentation, which breaks down ecosystem connectivity and produces isolated islands of habitats (Honnay et al. 2002). An additional change may be in timing, specifically in the advancement of spring events in temperate ecosystems, which is occurring 2.3 days earlier per decade (Parmesan and Yohe 2003). In coastal areas, ecosystems and urban systems will need to adapt to increases in sea level. These changes may not occur uniformly across the globe, however, but may occur faster in coastal areas or near the poles where temperature changes are more rapid.

As species shift in range, timing, and composition, ecosystem function will also change. The potential degradation of ecosystem function will threaten the ecosystem services provided to human society by nature, resulting in increased economic costs for these services. The impacts of global change on humanity will be contingent on society's ability to adapt to this change. Such an adaptation may necessitate coordinated ecosystem management across geopolitical boundaries to minimize the global impacts of climate change—a significant challenge considering the uncertainties regarding the rate and magnitude of change that has created difficulty in garnering international support to curb climate change.

A proposed adaptation technique is active ecosystem management. For instance, humans can mediate natural CO_2 storage by speeding up the rate of sequestration and reducing the release of already stored carbon. This may be accomplished through reforestation efforts or by increasing the growth rate of forests and practicing no-till agriculture. Additionally, deepwater or geologic sequestration of CO_2 may provide an alternative means to reduce atmospheric greenhouse gas concentrations.

Making informed ecosystem management decisions regarding global climate change comes with many challenges. Uncertainty surrounds the rates and magnitudes of change, due in part to the availability of data sources and the fact that many observations cannot be tied to a particular sampling station (Berliner 2003). Furthermore, considerable uncertainty arises from the fact that scientists have not yet solidified their understanding of many of the driving forces behind climate change models. A further area of uncertainty arises from not knowing how humans will continue to impact the Earth. Future use of fossil fuels, land-use change, and population increases are all variables, and they are largely dependent on societal decisions.

Recent climate change has the capability to dramatically alter the state and functioning of natural ecosystems. Much more must be learned quickly about the functioning of and the change in these systems in order to mitigate damage. By improving our understanding of natural ecosystems and their links to climate, we can be better prepared to properly manage our resources for future generations.

Charles E. FLOWER, Douglas J. LYNCH, and
Miquel A. GONZALEZ-MELER
University of Illinois at Chicago

See also in the *Berkshire Encyclopedia of Sustainability* Biodiversity; Biodiversity Hotspots; Carrying Capacity; Coastal Management; Complexity Theory; Ecological Forecasting; Food Webs; Human Ecology; Nitrogen Saturation; Regime Shifts; Resilience; Safe Minimum Standard (SMS)

FURTHER READING

Anderson, David E.; Goudie, Andrew S.; & Parker, Adrian G. (2007). *Global environments through the Quaternary*. New York: Oxford University Press.

Berliner, L. Mark. (2003). Uncertainty and climate change. *Statistical Science, 16,* 430–435.

Broecker, Wallace. (1997). Thermohaline circulation, the Achilles heel of our climate system: Will man-made CO_2 upset the current balance? *Science, 278,* 1582–1588.

Foley, Jonathan; Kutzbach, John; Coe, Michael; & Levis, Samuel. (1994). Feedbacks between climate and boreal forests during the Holocene epoch. *Nature, 371,* 52–44.

Gerlach, Terrence. (1991). Present-day CO_2 emissions from volcanoes. *Eos, Transactions, American Geophysical Union, 72*(23), 249, 254–255.

Graf, H.-F.; Feichter, J.; & Langmann, B. (1997). Volcanic sulfur emissions: Estimates of source strength and its contribution to the global sulfate distribution. *Journal of Geophysical Research, 102,* 10727–10738.

Hansen, James. (2009). *Storms of my grandchildren*. London: Bloomsbury.

Honnay, Olivier, et al. (2002). Possible effects of habitat fragmentation and climate change on the range of forest plant species. *Ecology Letters, 5,* 525–530.

Imbrie, John, & Imbrie, Katherine. (1979). *Ice ages: Solving the mystery*. Cambridge, MA: Harvard University Press.

Imbrie, John, et al. (1992). On the structure and origin of major glaciation cycles 1. Linear responses to Milankovitch forcing. *Paleoceanography, 7,* 701–738.

Intergovernmental Panel on Climate Change (IPCC). (2007). *Climate change 2007: The physical science basis. Contribution of Working Group I to the Fourth Assessment Report of the Intergovernmental Panel on Climate Change*. Cambridge, UK: Cambridge University Press.

Kaser, G.; Cogley, J. G.; Dyurgerov, M. B.; Meier, M. F.; & Ohmura, A. (2006). Mass balance of glaciers and ice caps: Consensus estimates for 1961–2004. *Geophysical Research Letters, 33,* L19501.

Lashof, Daniel, & Ahuja, Dilip. (1990). Relative contributions of greenhouse gas emissions to global warming. *Nature, 344,* 529–531.

Lean, Judith. (2010). Cycles and trends in solar irradiance and climate. *Wiley Interdisciplinary Reviews: Climate Change, 1,* 111–122.

Parmesan, Camill, & Yohe, Gary. (2003). A globally coherent fingerprint of climate change impacts across natural systems. *Nature, 421,* 37–42.

Petit, J. R., et al. (1999). Climate and atmospheric history of the past 420,000 years from the Vostok ice core, Antarctica. *Nature, 399,* 429–436.

Pinto, J. R.; Turco, R. P.; & Toon, O. B. (1989). Self-limiting physical and chemical effects in volcanic eruption clouds. *Journal of Geophysical Research, 94,* 11165–11174.

Prentice, I. C., et al. (2001). The carbon cycle and atmospheric carbon dioxide. In J. T. Houghton et al. (Eds.), *Climate change 2001: The scientific basis. Contributions of Working Group I to the Third Assessment Report of the Intergovernmental Panel on Climate Change* (pp. 185–237). Cambridge, UK: Cambridge University Press.

Ramanthan, Veerabhadran, et al. (1989). Cloud-radiative forcing and climate: Results from the Earth radiation budget experiment. *Science, 243,* 57–63.

Stocker, Thomas, & Schmittner, Andreas. (1997). Influence of carbon dioxide emission rates on the stability of the thermohaline circulation. *Nature, 388,* 862–865.

Vitousek, Peter; Ehrlich, Paul; Ehrlich, Anne; & Matson, Pamela. (1986). Human appropriation of the products of photosynthesis. *Bioscience, 36,* 368–373.

Vitousek, Peter; Mooney, Harold; Lubchenco, Jane; & Melillo, Jerry. (1997). Human domination of Earth's ecosystems. *Science, 277,* 494–499.

Wildung, Raymond; Garland, Thomas; & Buschbom, R. (1975). The interdependent effects of soil temperature and water content on soil respiration rate and plant root decomposition in arid grassland soils. *Soil Biology and Biochemistry, 7,* 373–378.

Grasslands

Grasslands are meadows and herbaceous expanses that may include scattered shrubs and trees. Characterized by pastoral cultures and grazing herbivores, grasslands are threatened by expanding cultivation, development, land fragmentation, and soil loss exacerbated by climate change and overexploitation. Arid grassland vegetation is largely controlled by factors beyond management influence, most notably rainfall and soils, which constrain strategies for sustainable use.

Grasslands are distinguished by short-statured vegetation allowing a clear view, a structure that has profound implications for grassland life. The dominant vegetation is commonly the grasses, or sometimes the grasslike sedges and rushes, often intermixed with colorfully flowering broad-leaved herbaceous plants and scattered trees and shrubs. Savannas, open shrublands, and tundra may be classified with grasslands for some purposes. Extensive grasslands show up on a global map (see figure 1), while smaller meadows and grassy openings are scattered throughout woodlands, forests, shrublands, and deserts. Grasslands may last for only a few years following fire or field abandonment, or persist for thousands of years.

Extensive grasslands are found in conditions as disparate as those of the high elevation Tibet-Qinghai plateau, the temperate Argentine pampas, the subalpine New Zealand tussock lands, Mediterranean basin scrub, and the tropical African savanna. Depending on the location, there is scientific debate about what causes persistent grasslands. Grasslands may result from human-set fires, cultivation, tree harvest, or pasture clearing, but they may also be a function of soils, aridity, irregular rainfall, extreme temperatures, high elevation, frequent fire, herbivory, or more likely some combination of factors. Excluding Antarctica and Greenland, about 41 percent of the Earth's surface is extensive grassland, and grasslands store up to 30 percent of the world's soil carbon, though such estimates vary widely according to the way grasslands are defined (White, Murray, and Rohweder 2000).

Denizens and Dependents

Grasses evolved more than 55 million years ago and spread rapidly in the Miocene, when the development of mountain ranges led to the creation of rain shadow areas with drier climates. The expansion of grasslands encouraged the rapid evolution of diverse herbivores. Grassland grazers have teeth and digestive systems able to cope with low-protein, fibrous, silica-containing grasses; they have eyes on the sides of their heads that maximize peripheral vision; and they have hooves that allow them to run swiftly across the open landscape to escape predators. Ruminants—cloven-hoofed grazers—have a four-chambered digestive system that can maximize the use of cell-wall fiber for energy. The diverse grazing fauna ranges from microbes and insects to wildebeests and giraffes.

Grassland plants usually are resilient or resistant to grazing. Some are toxic or taste bad, and some have unpalatable stickers that also help seeds hitch a ride on fur and clothing. Grasses usually have their growing points close to the ground, where they are missed by teeth or the blades of a mower, allowing rapid regrowth after leaf tissue is eaten or cut. Rhizomes or stolons that creep along the ground or through the soil allow the "sod-forming" grasses to spread without flowering. Deep persistent root systems allow "bunch" or "tussock" grasses to store nutrients and draw water. Annual grasses produce abundant seed that may persist for years in the soil. Many of these characteristics also help grasses survive or evade drought and fire.

Herds of grazing animals are sustenance to pack-hunting predators, such as lions, coyotes, and wolves, which exploit the ability to see their prey and their pack mates

Figure 1

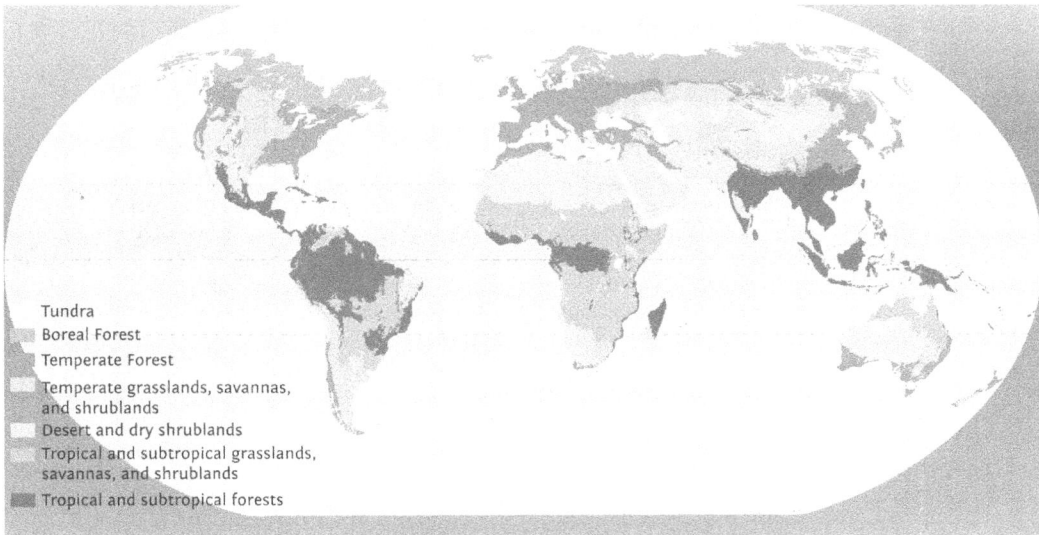

Tundra
Boreal Forest
Temperate Forest
Temperate grasslands, savannas, and shrublands
Desert and dry shrublands
Tropical and subtropical grasslands, savannas, and shrublands
Tropical and subtropical forests

Source: Adapted from Olson et al. (2001), Riccardo Pravettoni (cartographer), and UNEP/GRID-Arendal (2009).

What is defined as or included in grasslands varies, and grasslands themselves change over time as deserts increase or contract, and as woody vegetation invades or retreats. Indigenous grassland wildlife and cultures rely on mobility and flexibility to cope with climatic and resource heterogeneity and change. In addition to these large mapped areas, grasslands may exist as small areas scattered within other vegetation types.

across a distance. The fastest land animal on Earth—the cheetah—hunts in the grasslands and savanna of Africa. Eagles and hawks make use of open lines of sight to dive down on prey. Unique grassland birds nest in dense grasses or live in burrows. Reptiles feed on abundant rodents and insects in an ecosystem where primary productivity is concentrated close to the ground. In 2000 at least 17 percent of the centers of plant diversity (CPD) designated by the International Union for Conservation of Nature (IUCN)–World Conservation Union and World Wildlife Fund were found in grassland areas, with an additional 30 percent containing some grassland habitat. Of endemic bird areas (EBA) identified by Endemic BirdLife International, 11 percent have grasslands as key habitat. Grasslands are 26 percent of the terrestrial ecosystems targeted by World Wildlife Fund as outstandingly diverse and a priority for conservation (White, Murray, and Rohweder 2000).

Traditional Grassland Use

Pastoral cultures are characteristic of extensive grasslands, relying on herbivores to obtain high quality protein—blood, meat, and milk—from plants inedible by humans. Wild grazers move to water and rainfall, and migrate with the seasons, making use of the heterogeneity of grassland resources and avoiding severe drought and cold. While

Native Americans of the North American Great Plains did not domesticate the bison, they hunted and followed them across the prairies. Mongolian herders, African Masai, Sami reindeer herders, European shepherds, and many other pastoralists traditionally move with their herds. Mobility—of livestock and people—is a characteristic of grassland lifestyles worldwide and is based on use of shared grasslands that is moderated by tenure arrangements that tend to provide for flexibility in grassland use. Among traditional controls on the number and pattern of grazing animals are droughts that reduce numbers, ownership of wells by clans or tribes, spiritual traditions, and community governance that allocates grazing rights.

Pastoralists may maintain, expand, or create grasslands. Herders and hunting people set fires to clear shrubs and prevent invasion of woody plants into grasslands, meadows, and pastures. Where livestock rearing is part of a farming economy, cultivated forages limit the need for mobility for at least part of the year. In arid and tropical grasslands, however, cultivation can be damaging to soils and divert water from the ecosystem.

Threats to Sustainability

Well-watered temperate grasslands typically develop fertile, thick, organic upper layers of soil, making them attractive for farming. The prairies of North America, like temperate grasslands in many other areas, have been nearly completely converted to crop production. Some of these crops are domesticated grasses, or cereals. Three of these cereals—rice, wheat, and maize (corn)—have been estimated to produce more than half of all calories and the major sources of carbohydrates consumed by humans.

Arid and semiarid grasslands may have sandy soils that are highly vulnerable to erosion when vegetation is lost or reduced. The role of livestock grazing and pastoralism in desertification is a common concern. *Desertification* is a popular term for a posited cycle of increasing erosion and dryness caused by the loss of vegetative cover on desert soils. Grazing has been reduced, and pastoralists have been relocated or excluded from many areas where desertification has been attributed to overgrazing. Studies in the Sahara, however, have indicated that desert expansion and retraction is best predicted by changes in weather patterns or climate. The interrelationship with grazing is a topic of debate, and outcomes for soil carbon storage are part of this discussion. There is concern that climate change may lead to increases in aridity that favor desert spread and soil loss.

In native tropical and subtropical African savannas, the balance between woody vegetation and herbaceous plants is maintained by fire and the grazing of the world's greatest diversity of hoofed grazers. Cycles of flooding or drought may also limit woody plants. Tropical grassland soils, like tropical forest soils, are typically highly leached by heavy rainfall, moderate to poor in nutrients, and low in organic matter. Cultivation can leave the land infertile and is a major problem. Other threats include water diversion, overhunting, political upheaval, changes in fire frequency, and fragmentation of migration corridors. Deforestation may be caused by poor farmers and migrants clearing trees to increase pasture and farmland, as well as by large agricultural and livestock enterprises supported by distant investors.

Invasion of woody species into grasslands changes the habitat profoundly. Some scientists believe that increases in atmospheric carbon dioxide are contributing to the invasion of grasslands by mesquite in the southwestern United States. The exhaust from vehicles includes nitrogen, which increases the fertility of the soil near roadways. In the western United States this increased nitrogen supply was found to favor nonnative species. Invasive species may be brought in intentionally to increase forage production and for ornamental plantings, or accidentally with seed, livestock, and human visitors. Plowing of arid grasslands, sometimes referred to as *sod-busting*, led to the period of severe wind erosion known as the Dust Bowl in North America during the 1930s. Grasslands have been flooded, mined, and dried by groundwater pumping and water diversion. At the edges of deserts,

drifting sands and shifts in patterns of aridity may reduce or alter grasslands. Roads, housing developments, and changing land uses are today significant causes of grassland fragmentation and loss.

The migratory herds of the grasslands regularly conflict with sedentary lifeways, fixed ownerships, and hardened political boundaries. For example, in the United States, bison roaming beyond the boundaries of Yellowstone National Park run afoul of private landowners. The use of fire is constrained as a danger to crops and homes worldwide. In Europe, traditional transhumance—the herding of sheep, goats, and cattle into the mountains in the summer—has declined in the European Alps as young people seek alternative careers and development fragments the landscape, leading to brush invasion and biodiversity loss. United States ranchers often state that what they see as needlessly burdensome regulations, as well as landscape fragmentation and development, contribute to ranch sales for conversion to residential use. In the A-la Shan desert of northern China, herders are offered incentives to move to town, and large areas are fenced for grazing exclusion. Political upheaval, population shifts, urban expansion, and the destruction of traditional pastoral systems may lead to soil erosion and vegetation change, and the loss of grassland habitats.

Pastoralists, Grasslands, and Development

Chronic poverty and declining grassland productivity plague many grassland areas today. In response, almost all pastoral development projects have adopted what is referred to as the SIP model, in which sedenterization (S) leads to intensification (I) that then leads to increased agricultural production (P). This approach has not been successful.

Predicated on the assumption that the traditional institutions that sustained some pastoral systems for millennia have broken down, pastoral development programs generally introduce new institutions, including individuation of land ownership and establishment of a set *livestock carrying capacity* in an attempt to improve grasslands and incomes. Part of this is also sedentarizing pastoral people on their individuated parcels and encouraging them to intensify production with improved livestock breeds and feeds. Mobility becomes impossible, and no matter what the weather, herds stay put. Moreover, when households share grasslands they are able to share labor and separate species, breeding stock, young animals, males, females and so on, grazing them in areas most suited to each. When each household is restricted to one parcel, animals must be grazed together and by family rather than shared labor. Development to bring water to each individuated parcel may also become necessary, and construction of fences is a must. Unfortunately "improved breeds" and "intensive production" may require a higher level of nutrition and care than is possible for herders. This may lead to the cultivation of feed crops, disrupting the soil and requiring the use of scant water for irrigation. Traditionally, pastoralists graze mixed herds, for example, sheep, camels and goats, maximizing the ability to use diverse grassland resources and herd resilience to drought and disease. Intensification often calls for specialization to one species and even a single breed, increasing the risk of catastrophic failure.

Equilibrium-based ecological models for grasslands are based on the idea that plant competition leads eventually to a stable, *equilibrial (*or *climax)* plant community.

Where grasslands fit this model, a relatively constant number of livestock (or wild grazers) per unit area may be maintained. Disequilibrium-based ecological models seem to better explain the dynamics of arid and semiarid grasslands, or other grasslands where *disturbances,* including drought, erratic rainfall patterns, and fire, are common. In these areas, such abiotic factors may have greater influence on vegetation than grazing or biotic interactions like plant competition. Adaptability must be a core part of management strategies, including when necessary, herd reduction, movement, forage exchange, or alternate sources of forage. The two ways of understanding grassland ecosystems have different implications for grassland ecology, grazing systems, and development approaches. (See table 1).

TABLE I. Distinguishing Characteristics of Equilibrium and Disequilibrium Systems and Corresponding Strategies for Sustainable Use

Equilibrium System	Disequilibrium System
Ecological Assumptions	
Climate stability	Climate variability
Predictable primary productivity	Variable primary productivity
Livestock population controlled by density-dependent factors, i.e., biotic factors like vegetation	Livestock population controlled by density-independent factors, i.e., abiotic factors including rainfall and temperature.
Vegetation-livestock relationship is linear and reversible; change in stocking density creates predictable changes in plant composition	Vegetation-livestock relationship is nonlinear and nonreversible; livestock track unpredictable forage production
Potential carrying capacity can be predictable, and stocking density can be regulated according to potential carrying capacity	Prediction of potential carrying capacity not useful; uncertainties dominate future events
Ecosystem health is judged according to distance from an ideal, predictable, unchanging "climax"	Desirable plant communities are fostered within existing abiotically determined conditions
Management Implications	
Management oriented toward a single species of livestock	Management of multispecies herds: sheep, goats, camels, cattle, and horses
Strict regulation of stocking rates	Opportunistic stocking that changes according to existing conditions
Fixed tenure regimes (privatization or exclusive communal): focus on guarantee of ownership rights	Flexible tenure (complex mix of overlapping and integrated regimes): focus more on guarantee of access to resources
Cope with drought: stay in home grassland to resist disaster through infrastructure improvement and buying forage	Cope with drought: move to better grassland to avoid disaster
Conventional grassland management: forage farming, rotational grazing, paddock raising, and fencing	Opportunistic land use: mobility and flexibility, no fences

(*continued*)

TABLE 1. *Continued*

Equilibrium System	Disequilibrium System
Management Implications	
Blueprint development planning	Flexible, adaptive planning, with local involvement and a recognition of uncertainty
Management to increase carbon capture may be possible	Protecting soils will help prevent carbon loss, impossible to consistently increase carbon capture through management
Development Implications	
Economic goals are commercial, focus on commodity production: livestock development	Mostly subsistence economy, focus on livelihoods: pastoral development
Benefits are in financial terms, influenced by market	Benefits are in productive capital, not closely influenced by market
Service delivery package through centralized extension services; extension worker for technical delivery	Pastoral organizations for local management issues; extension workers as institutional organizers and mediators with outside world

Source: Adapted from Oba, Stenseth, and Lusigi (2000) and Scoones (1994).

Most grasslands are best understood as disequilibrium systems, where nonbiological (i.e., abiotic) factors like rainfall commonly overshadow the effects of management strategies by humans. In grasslands that require moderate moisture, however, equilibrium-based models may better explain ecological dynamics.

Improving Grassland Sustainability

In pastoral systems, innovative institutional arrangements that restore mobility have been seen. For example, in China some herders are voluntarily managing as collective units. In the United States and Australia, pastoralists are sharing or leasing rangelands to increase the ability to respond to drought with movement, and experimenting with grass banks—spare, shared pastures—that can be turned to during drought. In Africa, community-based natural resource management programs and safari enterprises that strengthen the value of wildlife and grasslands to communities have had some success in encouraging the conservation of elephants and other species. Consumer interest in "natural" meats from grasslands may help bolster pastoral economies where such markets exist. Strategies for increasing the flexibility of land use and the ability to respond quickly to changes in conditions are needed to maintain the relationship between grasslands and herders, and wild herbivores and grasslands.

Using grasslands to produce biofuels has been suggested and even incentivized in some countries. While this may increase the value of grasslands to the global economy, it also is possible that more grasslands will be plowed and planted with exotics. Emphasis on trees as carbon storage, and incentives for afforestation may also lead to the planting of grassland with trees and the encouragement of woody plants. Drought and fire may prove insurmountable obstacles to building carbon stocks this way. On the other hand, management practices that protect the soil and build it, where possible, have the potential to maintain and, in some cases where aridity or other factors external to management are not constraining, to increase the contribution of grasslands to global carbon storage.

In the coming decades species and cultures relying on access to large-scale open grasslands will lose ground. Poverty and political upheaval will continue to destabilize grassland systems. Climate shifts will also affect grasslands, and an increasingly limited capacity to be flexible in land use and grazing patterns will conflict with these challenges.

Lynn HUNTSINGER
University of California, Berkeley

LI Wenjun
University of Peking

See also in the *Berkshire Encyclopedia of Sustainability* Animal Husbandry; Bioenergy and Biofuels; Ranching

FURTHER READING

Adler, Paul. R.; Sanderson, Matt A.; Weimer, Paul J.; & Vogel, Kenneth P. (2009). Plant species composition and biofuel yields of conservation grasslands. *Ecological Applications, 19*(8), 2202–2209.

Behnke, Roy H.; Scoones, Ian; & Kerven, Carol. (1993). *Range ecology at disequilibrium: New models of natural variability and pastoral adaptation in African savannas.* Regent's College, London: Overseas Development Institute.

Christensen, Lindsay; BurnSilver, Shauna; & Coughenour, Michael. 2006. Integrated assessment of the dynamics, stability, and resilience of the Inner Mongolian grazing ecosystems. *Nomadic Peoples, 9*(1 & 2), 131–145.

Ellis, Jim E., & Swift, Dave M. (1988). Stability of African pastoral ecosystems—Alternate paradigms and implications for development. *Journal of Range Management, 41*(6), 450–459.

Fernandez-Gimenez, Maria E., & Allen-Diaz, Barbara. (1999). Testing a non-equilibrium model of rangeland vegetation dynamics in Mongolia. *Journal of Applied Ecology, 36*(6), 871–885.

Gillson, Lindsay, & Hoffman, Timm M. (2007). Rangeland ecology in a changing world. *Science, 315*(5808), 53–54.

McAllister, Ryan R. J. (2010). Livestock mobility in arid and semi-arid Australia: Escaping variability in space. *Pastoralism: Research, Policy, and Practice, 1*(1), 38–55.

McIntyre, Sue; Heard, Katina M.; & Martin, Tara G. (2003). The relative importance of cattle grazing in subtropical grasslands: Does it reduce or enhance plant biodiversity? *Journal of Applied Ecology, 40*(3), 445–457.

Neely, Constance; Bunning, Sally; & Wilkes, Andreas. (2009). *Review of evidence on drylands pastoral systems and climate change.* Retrieved January 20, 2010, from ftp://ftp.fao.org/docrep/fao/012/i1135e/i1135e00.pdf

Oba, Gufu; Stenseth, Niles C.; & Lusigi, Walter J. (2000). New perspectives on sustainable grazing management in arid zones of sub-Saharan Africa. *BioScience, 50*(1), 35–51.

Olson, David M., et al. (2001). Terrestrial ecoregions of the worlds: A new map of life on Earth. *Bioscience, 51*(11), 933–938.

Scoones, Ian. (1994). *Living with uncertainty: New directions in pastoral development in Africa.* London: Intermediate Technology Publications.

Seaquist, J. W.; Hickler, Thomas; Eklundh, Lars; Ardo, Jonas; & Heumann, Benjamin W. (2009). Disentangling the effects of climate and people on Sahel vegetation dynamics. *Biogeosciences, 6*(3), 469–477.

Silver, Whendee L.; Ryals, Rebecca; & Eviner, Valerie. (2010). Soil carbon pools in California's annual grassland ecosystems. *Rangeland Ecology and Management, 63*(1), 128–136.

Tucker, Compton J., & Nicholson, Sharon E. (1999). Variations in the size of the Sahara Desert from 1980 to 1997. *Ambio, 28*(7), 587–591.

United Nations Environmental Programme (UNEP) / GRID-Arendal. (2008). Savannas and tropical grasslands. Retrieved Jan. 20, 2010, from http://maps.grida.no/go/graphic/savannas-and-tropical-grasslands

Wedin, Walter F., & Fales, Steven L. (2009). *Grassland: Quietness and strength for a new American agriculture.* Madison, WI: American Society of Agronomy.

Weiss, Stuart B. (1999). Cars, cows, and checkerspot butterflies: Nitrogen deposition and management of nutrient-poor grasslands for a threatened species. *Conservation Biology, 13*, 1476–1486.

Westoby, Mark; Walker, Brian H.; & Noy-Meir, Immanuel. (1989). Opportunistic management for rangelands not at equilibrium. *Journal of Range Management, 42*(4), 266–274.

White, Robin; Murray, Siobhan; & Rohweder, Mark. (2000). *Pilot analysis of global ecosystems (PAGE): Grassland ecosystems.* Washington, DC: World Resources Institute. Retrieved January 20, 2010 from http://www.wri.org/publication/pilot-analysis-global-ecosystems-grassland-ecosystems

Greenhouse Gases

Beginning with the Industrial Revolution in the eighteenth century, human activity has contributed to the increase of greenhouse gases in the atmosphere. This changing atmospheric balance has led to the warming of the planet. Scientists understand the relation of each of the known greenhouse gases with the Earth as a system and with human activity.

Greenhouse gases are gases in the atmosphere that absorb electromagnetic radiation in the thermal (infrared) energy range. Solar radiation, which is the primary source of energy for Earth, is high-energy radiation (visible and ultraviolet), and a major proportion of the visible radiation is transmitted through the atmosphere and absorbed by the Earth. In return the Earth reemits the radiation as a variety of energies, usually of lower energy value (thermal energy). The greenhouse gases that are in the atmosphere can easily absorb the thermal energy radiation that is emitted back into the atmosphere from Earth. This is much like what happens in a greenhouse, where the glass or plastic allows the solar energy to be transmitted, and the reemitted thermal energy is absorbed inside the greenhouse, thus heating the greenhouse. The most abundant greenhouse gases in our atmosphere are water vapor, carbon dioxide, methane, and nitrogen oxides. Others include ozone, hydrofluorocarbons, and many complex human-made gases.

Greenhouse Gas Concentrations

The amount of each of the most abundant greenhouse gases present in the atmosphere is related to the natural environment and to the activities of humans. There is strong evidence in modern times of a direct correlation between greenhouse gas concentrations in the atmosphere

and global temperatures. The spike in the amount of carbon dioxide, methane, and nitrogen oxide in the atmosphere over the last two hundred years is extraordinary.

Water Vapor

Water vapor is the most abundant of all of the greenhouse gases. Its abundance does not show the same correlation with the effects of the Industrial Revolution as the other greenhouse gases, and its role in climate change is also the least understood, because there is no historic data, as there is for the other greenhouse gases, so it is not possible to deduce the human effects on its concentrations. The amount of water vapor is several orders of magnitude greater then the other greenhouse gases, and it varies greatly from one location to another. Although as a substance, water is one of the most studied, its relation to climate change and global warming is less understood. The abundance of water vapor is related to temperature and is expected to increase as global atmospheric temperatures increase. The direct effect should be an increase in the greenhouse effect, thereby amplifying warming. On the other hand, as temperatures rise and the amount of water vapor increases, there is an increased probability of cloud formation and thus cloud cover, which will have the affect of shielding the Earth from incoming radiation and thus reduce warming. The balance between the two effects is not well understood.

Carbon Dioxide

The most investigated of the greenhouse gases is carbon dioxide. The increase in the abundance of carbon dioxide can be traced back to the beginning of the Industrial Revolution in the eighteenth century and the combustion of carbon fuels. The unit most used for expressing carbon dioxide concentrations is ppm (parts per million). For at least ten thousand years prior to the Industrial Revolution, there was consistently 260–280 ppm of carbon dioxide, compared to today's carbon dioxide concentration of approximately 400 ppm (Ahn et al. 2004; Petit et al. 1999; Siegenthaler et al. 2005). Some scientists have identified concentration of around 350 ppm as the tipping point, a point of no return at which catastrophic changes to the planet are inevitable. The most noted change is global warming: the melting of the polar ice caps will result in significant sea-level rises and the consequent destruction of coastal habitats.

The increase of carbon dioxide in the atmosphere is the result of two factors: increased combustion of carbon fuels and the decrease in the planet's ability to remove carbon dioxide. Carbon dioxide is removed from the atmosphere either by being absorbed by the oceans or by the carbon dioxide cycle, the conversion of carbon dioxide into oxygen during photosynthesis. As the oceans absorb more and more carbon dioxide, their capacity is becoming limited, although the extent of this limitation is unknown. The oceans are also becoming more acidic because of the increased absorption of carbon dioxide. At the same time, rapid deforestation since the 1960s has diminished the effectiveness of the carbon dioxide cycle to reduce carbon dioxide concentrations in the atmosphere.

Methane

Methane is approximately twenty times more effective than carbon dioxide as a greenhouse gas, but its atmospheric concentration is two hundred times less. Much of the methane in the atmosphere is from human activity, agriculture in particular. A recent EPA report estimates more than 50 percent of global methane emissions are human related (US EPA 2010). In the future, the increasing melting of the globe's tundra regions, as a result of global warming, will release additional methane on thawing. Methane is formed by the decay of biomass that, in turn, is trapped both on land and in the sea in the Earth's permafrost, which is melting at increasing rates.

Human Activity and Greenhouse Gas Trends

Much human economic activity today adds greenhouse gases to the atmosphere. The economic sectors that have the greatest impact include agriculture, commerce, housing, transportation, and industry. Humans account for an annual increase of 2 ppm in the overall amount of carbon dioxide (NOAA 2010). Concentrations of the other greenhouse gases also continue to rise, despite growing efforts to control the amount of greenhouse gases humans are releasing into the atmosphere.

John G. STEVENS
University of North Carolina Asheville

See also in the *Berkshire Encyclopedia of Sustainability* Carbon Capture and Sequestration; Coal; Glaciers; Heating and Cooling; Hydrogen Fuel; Petroleum; Oceans and Seas

FURTHER READING

Ahn, Jinho, et al. (2004). A record of atmospheric CO_2 during the last 40,000 years from the Siple Dome, Antarctica ice core. *Journal of Geophysical Research, 109*, D13305.

Alley, Richard B. (2002). *The two-mile time machine: Ice cores, abrupt climate change, and our future.* Princeton, NJ: Princeton University Press.

Holdren, John (2008). Meeting the climate change challenge. Eighth Annual John H. Chafee Memorial Lecture on Science and the Environment. Washington, DC: National Council for Science and the Environment.

Intergovernmental Panel on Climate Change (IPCC). (2006). 2006 guidelines for national gas inventories. Retrieved June 20, 2011, from http://www.ipcc-nggip.iges.or.jp/public/2006gl/index.html

National Climatic Data Center (n.d.). Greenhouse gases: Frequently asked questions. Retrieved June 20, 2011, from http://www.ncdc.noaa.gov/oa/climate/gases.html

National Oceanic and Atmospheric Administration (NOAA). (2010a). Emissions of potent greenhouse gas increase despite reduction efforts. US Department of Commerce. Retrieved June 20, 2011, from http://www.noaanews.noaa.gov/stories2010/20100127_greenhousegas.html

National Oceanic and Atmospheric Administration (NOAA). (2010b). NOAA annual greenhouse gas index (AGGI). Retrieved June 20, 2011, from http://www.cmdl.noaa.gov/aggi

National Oceanic & Atmospheric Administration (NOAA) Earth System Research Laboratory. (2011). Trends in atmospheric carbon dioxide [Data file]. Retrieved September 20, 2011, from ftp://ftp.cmdl.noaa.gov/ccg/co2/trends/co2_mm_mlo.txt

Petit, Jean Robert, et al. (1999). Climate and atmospheric history of the past 420,000 years from the Vostok ice core, Antarctica. *Nature, 399*(6735), 429–436.

Siegenthaler, Urs, et al. (2005). Stable carbon cycle: Climate relationship during the Late Pleistocene. *Science, 310*(5752), 1313–1317.

United States Environmental Protection Agency (EPA). (2010, April). *Methane and nitrous oxide emissions from natural sources.* Retrieved September 20, 2011 from http://epa.gov/methane/pdfs/Methane-and-Nitrous-Oxide-Emissions-From-Natural-Sources.pdf

US Environmental Protection Agency (EPA). (2011). 2011 inventory of greenhouse gas emissions and sinks. Retrieved June 20, 2011, from http://www.epa.gov/climate change/emissions/index.html

Insects

While most people may consider insects little more than useless nuisances, insects have benefited humankind in many ways for thousands of years: for sustenance, in manufacturing, for medical use, in agriculture, and in the control of problem pests. They also have some appeal as tourist attractions and can be useful indicators of environmental health and well-being.

Throughout history insects have provided significant natural resources for humans. These have included species that were eaten by aboriginal peoples for subsistence or to make medicines and poisons. Lac insects produce a resinous secretion that was used to make lacquer. Cochineal insects were harvested for their red pigment. In the developed world and in Western culture, direct insect utilization is probably now at an all-time low, with only two insects carrying any serious burden—the honeybee and the silkworm, which have been subsumed into an industrialized and multinational global economy. Nevertheless, insects are still important for human well-being through their sometimes hidden roles as pollinators, biocontrol agents, and environmental indicators, helping to maintain balance in the natural world.

Eating Insects

The Bible goes to some lengths to suggest a scheme of nutritional balance, listing animals that can and cannot be safely eaten. Listed along with the usual cloven-hoofed, cud-chewing, meat providers are several insects: "Even these of them ye may eat; the locust after his kind, and the bald locust after his kind, and the beetle after his kind, and the grasshopper after his kind" (Leviticus 11:22). There has long been debate about exactly which species are implied, especially as the text seems to be

confused about how many legs these animals might have. (Four rather than six are suggested.)

It seems likely that some loss in translation has occurred through the centuries, but away from the eastern Mediterranean, locusts, grasshoppers, and their kin are still eaten, if not commonly, then at least widely. Other well-known examples of insects being eaten by humans include bogong moths (*Agrotis infusa*), collected by indigenous Australians in large numbers as they roost, and witchetty grubs (large wood-boring larvae of various moths, mainly *Endoxyla leucomochla*), also in Australia.

These insects may once have been staples in local diets throughout the world, but they are now reduced almost to being nutritional novelties on par with chocolate-coated ants or mint-flavored lollipops containing a grasshopper (marketed in the United States under the name Cricket Lick-it). The decline in aboriginal insect foods and the limited niche marketing of these quirky modern offerings appear to offer no threat to the sustainability of their use.

Insects in Medicine

Most traditional medicines have been herb, or at least plant, based and have led to important modern drugs like morphine (from poppies), aspirin (from sallow), and digitalis (from foxglove). Insect-based medicaments are few, although several have been known since ancient times. The corrosive and toxic oil cantharidin, secreted by the aptly named oil or blister beetles (family Meloidae), was once widely used to treat various complaints, such as wart and tattoo removal, to induce blistering (when this was considered medically useful), as an aphrodisiac, and as an executioner's poison. Traditional Eastern medicines also sometimes contain invertebrate ingredients, including centipedes, scorpions, mantises, cicadas, and crickets.

The most important insects in Western medicine today are flesh-fly maggots (*Lucilia* species). The discovery that fly maggots infesting battle wounds actually aid the recovery of the victim has been made independently over several centuries, including during the Franco-Habsburg War (1551–1559), Napoleon's Egyptian Campaign (1798–1801), the Crimean War (1853–1856), and the American Civil War (1861–1865) (Service 1996). It was noted that the flies visited the untended wounded on the battlefield, and during the days before these men were rescued or taken prisoner, the maggots fed in the open wounds, eating diseased and necrotic tissue in preference to healthy flesh. Sterile laboratory-bred fly maggots are now routinely used in hospitals around the world to clean necrotizing or even gangrenous and infected wounds after surgery or injury.

Commercial Products

Cochineal (*Dactylopius coccus*) is a small, aphid-like insect that feeds on various Central and South American prickly pear cactuses (genus *Opuntia*). It produces a bitter but very bright crimson body pigment called carmine. Although best known for coloring confectionary, cochineal was originally used to dye fabrics and cosmetics and has been traded internationally since at least 2200 BCE. In 1600, after the Spanish conquest of Mexico, cochineal was second only to silver in its European trade value. In 1787 the British Empire deliberately introduced cochineal-infested prickly pears to Australia to establish a source there for this important dye, which was used in the clothing industry and especially for British military uniforms. Unfortunately, the host cactuses became invasive weeds, clogging and degrading pasture until a biological solution was sought in the 1920s (see below). Meanwhile, with the advent of chemically synthesized dyes, the labor-intensive cochineal industry collapsed worldwide. More recently, however, as a result of toxicity scares, natural food colors like cochineal are making a comeback, and cochineal farming is being resumed on an increasing scale. In 2009, the US Department of Agriculture issued updated guidelines for labeling foods and medicines containing cochineal, often also called 'natural red' dye.

Lac insects, various genera of scale insect (Hemiptera, Coccoidea) found in China, India, Southeast Asia, and Mexico, are wingless relatives of greenfly (aphids) . They secrete a hard red resin that solidifies on, and is then harvested from, cut twigs and used as a lacquer. Traditionally used in the Far East on expensive decorative and ornamental wares, and in Europe as varnish on violins and other high quality merchandise, lac resin became a serious world commodity in the nineteenth century. Dissolved in methylated spirits or other solvents, it became shellac and was widely used to varnish floors, wood panels, and furniture as well as musical instruments. In the first decades of the twentieth century the resin, now simply called "lac," found use in small decorative moldings like miniature picture frames, jewelry, and dentures, but it is best remembered for early pressed phonograph records. Plastic and manufactured artificial resin varnishes have now become much more common, but shellac is still marketed as a varnish for its intrinsic strength, delicate tone, aesthetic qualities, and natural heritage.

Industrial Insects

Humans regularly exploit only two insects, but they do this on an industrial scale. Honeybees are almost wholly domesticated and housed in apiaries that range from single backyard hives run by hobby keepers to bee farms with hundreds of colonies. The main products are honey and wax, and economies of scale have brought the cost of these once luxury commodities down to the status of everyday items. Honeybees, particularly the western honeybee, *Apis mellifera*, can no longer really be regarded as wild, but there is some prospect of using less domesticated, sometimes feral, subspecies to reintroduce genetic hardiness at a time when industrialized uniformity is being partly blamed for honeybee declines.

Silkworms, more specifically the caterpillars of the silk moth, *Bombyx mori*, are also unknown in the wild, having been selectively bred by humans for thousands of years. Reared indoors on harvested leaves of white mulberry, each fully grown larva uses a single strand of silk to spin its cocoon before transforming into an adult moth. Boiling the silk in water breaks down one of the outer-layer proteins, and each strand can then be unraveled by machine before spinning, dying, and weaving.

Silk is still relatively labor intensive, but mechanization and huge economies of scale have also reduced it from the point where the one-time cloth of emperors is now virtually a commonplace fabric. Global silk production is now in the hundreds of thousands of tonnes annually. Before the advent of modern synthetic fibers, alternatives to *Bombyx* silk were sought. The closely related gypsy moth, *Lymantria dispar,* was deliberately brought to the United States in 1868 as part of attempts to breed new hybrid silk moths that would feed on locally available food plants. It soon escaped from the laboratory and has become a notorious forestry pest, defoliating millions of acres of trees each year.

Insect Tourism

Just as wildlife tours are popular with vacationers and travelers—big game safaris, dolphin encounters, and bird-watching cruises are already well established—there is potential for insect tourism. Butterfly enthusiasts already organize trips to observe the rich, insect faunas of both temperate and tropical wildlife habitats, and although these are usually small-scale operations, there are possibilities for mass tourism.

The glow-worm caves of Australia and New Zealand are established commercial tourist attractions. The glowing larvae of certain flies (*Arachnocampa* and other species) live on the roofs of large underground caverns. They feed on other insects that waft in on air and water currents and are then attracted upward by the shimmering lights, where they become entangled in sticky silk-thread traps created by the maggots. The spectacular, constellation-like displays are usually only visible on guided tours by boat or on foot, and some protection is offered by limiting the number of visitors.

The monarch butterflies, *Danaus plexipus,* of Canada and the United States make an annual migration to overwintering sites in about thirty sheltered valleys west of Mexico City. An estimated 250 million butterflies make the journey and arrive each November to settle in the oyamel fir trees and surrounding bushes. The spectacle of massive trees creaking under the weight of millions of butterflies, and the rustle of wings in air alive with insects is quite awe inspiring and already a regular pilgrimage site for naturalists, wildlife-film crews, photographers, and sightseers. This is a highly vulnerable habitat, and although at present not easily accessible, it could be upset by regular tourist traffic, whether by road, foot, or mule.

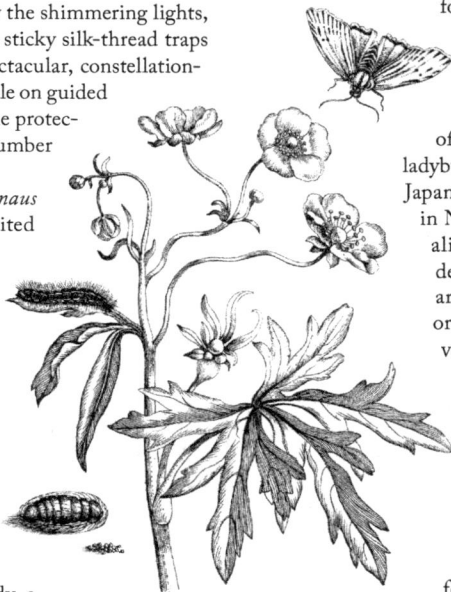

Biocontrol

Insects have proved very useful as biocontrol agents, redressing upsets in the balance of nature caused by accidental (and deliberate) releases of nuisance species, pests, and weeds around the globe. An early success was the 1888 introduction of the vedalia ladybug, *Rodolia cardinalis,* from Australia to combat the cottony cushion scale *Icerya purchasi,* a New Zealand insect that had been destroying citrus crops in California. A Mexican moth, *Cactoblastis cactorum,* was imported into Australia between 1926 and 1932 to control prickly pear cactuses. These had been brought from Central America in the nineteenth century to feed the newly established cochineal industry, but they had become an invasive weed. The moth's caterpillars fed on the prickly pears and soon kept them under control (Naumann et al. 1991).

Another notable success was the introduction of several species of African and American dung beetles into Australia in the middle of the twentieth century. Cattle dung, which is semiliquid, was not being removed and recycled by the native beetle species, which were adapted to living in dry, marsupial pellets. Accumulating dung was smothering the landscape and providing rich habitat for the Australian bush fly, *Musca vetustissima,* which reached plague proportions.

Such introductions can be dangerous, however. The release of the European seven-spotted ladybug (*Coccinella septempunctata*) and Japanese ladybug (*Harmonia axyridis*) in North America to combat various alien aphids has been blamed on declines of native ladybugs, which are outcompeted by the newcomers or even eaten by them and their voracious larvae.

A highly precise, though technologically complex, control process is offered by the sterile male technique. Mass release of irradiated—and thus sterile—male insects into a target area means that virtually all mating incidents with local females produce nonviable eggs, leading to population crash. When the flesh-eating South American screwworm fly, *Cochliomyia hominivorax,* was accidentally introduced into Libya in 1988, it soon infested the small cuts and wounds that farm animals regularly receive from barbed wire, thorn bushes, and each other, threatening a continent-wide epidemic in Africa and massive loss of animal (and human) life. At the time, Libya was a pariah state, but the United Nations initiated an eradication program to avert a plague of biblical proportions. Between December 1990 and October 1991, 1.3 billion sterile flies raised in the United States were dropped in small cardboard boxes from low-flying aircraft, and the screwworm fly was successfully eradicated (WHO 2011). Similar releases of sterile Mediterranean fruit flies (*Ceratitis capitata*) are used to control this almost cosmopolitan orchard pest in mainland United States.

Not all biocontrol organisms are exotic predators brought in to combat alien pests. There are large domestic markets for providing ladybugs and other predators for garden control. The convergent ladybug, *Hippodamia convergens,* is so called because every year it converges on the same rocky hollows in the hills and mountains of North America. Early arrivals give off a safety pheromone (scent) that recruits others to join, until clusters of many hundreds of thousands are knotted together. They are then harvested using bucket and spade and packaged for sale to gardeners. The pheromones linger and identify the same hollows to successive generations of ladybugs year after year.

Environmental Indicators

Popular environmental interest is usually focused on large, impressive, and beautiful animals and birds. While the occasional butterfly may be appreciated for its flower-decorating prettiness, most insects are seen as bird food or picnic nuisances, if they are considered at all. Nevertheless, insects remain one of the most important groups of organisms that directly benefit humans.

Their small size means that insects are seldom directly exploited by humans, but their huge numbers constantly bring them into hidden contact with all manner of human activities. The benefits usually ascribed to insects include pollination of plants (for both aesthetic flowers and productive crop fruits and vegetables); recycling of animal dung, carrion, fallen timber, and other decaying organic matter; food for other animals (insects dominate the middle portion of almost all food chains and webs); and maintaining the balance of nature as "helpful" insects kill and eat other pest species.

There is, however, a more subtle benefit offered by these diminutive creatures, and one that can be directly utilized by humans. Most insect species are confined by very precise and narrow habitat constraints; they may feed on a single plant species (or a small part of one plant species), live only in a certain microhabitat, or be limited to a precise geographical area. By studying the intricate ecologies of insects, entomologists are able to make broader assessments of habitat value (or loss), environmental health (or degradation), biodiversity, pollution, and climate change. Because of their small size, vast numbers, and broad diversity, insects can be used as measurable indicators of what is well, or unwell, in the world.

Richard A. JONES
Royal Entomological Society

See also in the *Berkshire Encyclopedia of Sustainability* Agriculture (*several articles*); Ecotourism; Honeybees; Indigenous and Traditional Resource Management; Insects—Pests; Medicinal Plants; Pest Management, Integrated (IPM)

FURTHER READING

Aston, David, & Bucknall, Sally. (2004). *Plants and honeybees: Their relationships.* Hebden Bridge, UK: Northern Bee Books.

Benjamin, Alison, & McCallum, Brian. (2008). *A world without bees.* London: Guardian Books.

Berenbaum, May R. (1995). *Bugs in the system: Insects and their impact on human affairs.* Cambridge, MA: Helix Books.

Chang, Franklin. (1982, September 15). Insects, poisons, and medicine: The other one percent (Proceedings from Presidential Address, presented at the December, 1979 meeting of the Hawaiian Entomological Society). *Hawaiian Entomological Society, 24*(1), 69–74. Retrieved January 3, 2011, from https://scholarspace.manoa.hawaii.edu/bitstream/10125/11139/1/24_69-74.pdf

Clarke, John. (1839). *A treatise on the mulberry tree and silkworm: And on the production and manufacture of silk.* Retrieved January 3, 2011, from http://www.biodiversitylibrary.org/item/82191

Greenfield, Amy Butler. (2005). *A perfect red: Empire, espionage, and the quest for the color of desire.* New York: Harper Collins Press.

Jones, Richard A. (2010). *Extreme insects.* London: Harper Collins.

Naumann, I. D., et al. (Eds.). (1991). *The insects of Australia: A textbook for students and research workers* (2nd ed.) (2 vols.). Carlton, Australia: Melbourne University Press, CSIRO, Division of Entomology.

Penang Butterfly Farm. (2011). Butterfly houses of the world. Retrieved January 3, 2011, from http://www.butterfly-insect.com/butterfly-houses-of-the-world.php

Scoble, Malcolm J. (1995). *The lepidoptera: Form, function and diversity* (2nd ed.). London: Oxford University Press & Natural History Museum.

Service, Mike W. (1996). *Medical entomology for students.* Cambridge, UK: Chapman & Hall.

World Health Organization (WHO). (2011). Homepage. Retrieved January 3, 2011, from http://www.who.int/en/

Landscape Architecture

Landscape architecture involves a wide variety of out-door design project types in both the natural and built environments, and as such includes making an ecological inventory and analysis to identify opportunities and constraints for land conservation and development. The US-based Sustainable Sites Initiative (SITES) exemplifies recent innovations being adapted in North America, northern and central Europe, Australia, South Korea, China, and Japan; these innovations are based on attention to the ways ecosystems function.

Landscape architecture is the art and science of arranging land so as to adapt it most conveniently, economically, functionally, and aesthetically to any of the varied desires of people. A landscape is the synthesis of all the natural and cultural features—fields, hills, forests, farms, deserts, water, and buildings or other structures—that distinguish one part of the surface of the Earth from another part. Landscape architecture, which involves the planning, design, and management of natural and built environments (Hooper 2007), positively contributes to ecosystem management approaches by incorporating such practices as effective drainage systems, self-sustaining vegetation and wildlife-supporting habitat, site planning to make the best use of solar energy, and resource recycling.

A Landscape Typology

Landscape architects are involved in a wide range of project types outdoors. William Tishler (1989), Stephen Carr et al. (1992), and Mark Francis (2001) have identified those project types, most of which are included in the following list. Frederick Steiner, a professor of landscape architecture and planning at University of Texas at Austin, expanded and/or consolidated the list, providing brief examples of some of the ways in which landscape architects contribute to sustainable ecosystem management:

- *Brownfield redevelopment:* finding sustainable designs for the reuse of industrial or commercial sites, many of which may be contaminated with chemicals and toxins.
- *Botanical gardens:* increasing the use of indigenous plants and sustainable irrigation practices.
- *Campuses:* adapting landscape planning to support sustainable practices in infrastructure development (e.g., using environmentally friendly materials and situating buildings to enhance the use of solar energy to reduce the carbon footprint); community building (providing an environment that contributes to human wellness and well-being); and learning (encouraging curriculum development in the field, which will foster public awareness and open opportunities for a new generation of landscape architects).
- *Cemeteries:* adapting principles (both aesthetic and functional) to enhance the spiritual connection of humans to the natural world and reduce the impact of human activity in it.
- *City, suburban, and rural town planning, as well as regional development:* supporting and contributing to the development of sustainable infrastructure (roadways, subways, and railways; municipal buildings; power and telephone lines; and the configuration of housing and commercial developments); and assessing sustainable land use (commercial, industrial, agricultural, and residential) as the focus or inspiration for landscape design.
- *Community open spaces (urban and rural), gardens (private and public), and plazas:* providing aesthetically pleasing and sustainability-focused environments to enhance people's quality of life.

- *Green roofs:* adapting previously or newly built roofs as gardens by implementing lightweight soils (enhanced with minerals or nonorganic fillers), climate-appropriate plants (including grasses and shrubs), root barriers, drainage layers, and waterproof membranes to protect the roof—thereby providing insulation, absorbing rainwater rather than creating runoffs, and benefiting birds and other wildlife.
- *Green walls:* using vegetation on walls to mitigate climate and to provide food and habitat for birds and reptiles.
- *Greenways (coined as a combination of the terms* greenbelt *and* parkway)*:* adapting former railroads, highways, or other transportation routes into a multipurpose "linear park" with vegetation (e.g., the High Line in New York City, the Gold Coast Oceanway in Australia, the EuroVelo cycles routes, and the Trans Canada Trail).
- *Historic landscapes:* using sustainable methods to conserve flora and fauna while preserving the aesthetic sense of the original architecture and design.
- *Housing environments:* planning or reassessing residential areas to increase sustainable practices, such as facilitating the use of solar power, providing open green spaces, and allowing access to public transportation.
- *Institutional and corporate landscapes:* finding innovative ways to redesign or renovate the spaces surrounding the architecture and infrastructure of preexisting (often dehumanizing and unsustainable) sites, and to utilize or adapt principles from other ecosystem friendly approaches when designing new sites.
- *National forests and parks, state parks, and other recreation areas:* working with federal, state, or municipal governments to make the best sustainable use of existing rules, regulations, and laws; and supporting newer and greener practices, including water, soil, and wildlife management.
- *Olympic, World's Fair, Expo, and other special-use venues:* upholding a long tradition of the landscape architect's role in developing special venues while focusing on newer and more sustainable building methods, energy use, and access to infrastructure (e.g., transportation).
- *Restored and/or reclaimed natural landscapes:* employing environmentally friendly methods for public access, such as boardwalks; removing invasive flora species, such as purple loosestrife and kudzu; and adapting principles of sustainable water management, such as rain gardens to control runoff.
- *Urban parks and playgrounds:* collaborating with practitioners in the field of urban forestry, which recognizes the benefits of using green spaces to combat air pollution, support biodiversity, and contribute to human emotional and physical well-being.
- *Waterfronts (and/or waterways):* using green practices to slow or prevent erosion and protect water quality, preserve and enhance the historic character and meaning of a waterfront landscape, and provide public access to waterfront areas.
- *Zoos:* providing environments that come as close as possible to the natural habitat of a species, and using sustainable practices (e.g., water management), to foster self-sustaining vegetation.

Two Landscape Architect Pioneers

The American landscape architect Frederick Law Olmsted Sr. (1822–1903), who designed Central Park in New York City in 1858, worked hard in collaboration with his colleague Calvert Vaux, his sons, John C. Olmsted and Frederick Law Olmsted Jr., and others (including Charles Eliot), to establish landscape architecture as a profession. The senior Olmsted was also involved in the design of private gardens (The Biltmore in Asheville, North Carolina), college campuses (Stanford University), metropolitan park and parkway systems (Boston, Louisville, and Buffalo), new community design (Riverside, Illinois), a World's Fair (the 1893 World's Columbian Exposition in Chicago), institutional landscapes (the New York Hospital for the Insane and the McLean Hospital grounds), and state parks (Niagara Falls State Park). Frederick Law Olmsted Jr. continued these activities with his brother and also played a leadership role in establishing the national parks system in the United States. Before the Olmsteds, traditional commissions for landscape gardeners in North America and Europe were for private clients. Olmsted and his followers changed this and, in effect, took the English picturesque tradition public.

Jens Jensen (1860–1951), a landscape architect born in Demark (but who practiced primarily in Chicago; Door County, Wisconsin; Dubuque, Iowa; and Springfield, Illinois), brought to his practice a personal belief in the renewing and civilizing powers of nature. As the leader of the Prairie Style of landscape architecture, he inspired a movement to conserve threatened scenic natural areas; he has been described as more devoted to the landscape of the Midwest than many born there (Henderson 1985). His acquaintances and supporters included the sociologist and reformer Jane Addams; the architect Frank Lloyd Wright; Harriet Monroe, an editor, poet, and patron of the arts; the botanist Henry Cowles; and Illinois governor Frank Lowden. In the twenty-first century the Jens Jensen Legacy Project, sponsored by the Chicago Department of Cultural Affairs and the Chicago Park District, seeks to provide educational opportunities for children and adults, to bolster current efforts to restore and preserve Jensen's projects, and to raise awareness of Jensen's Prairie Style to a new generation of landscape designers.

Landscape architecture, inspired in part by the early work of Olmsted and Jensen—and redefined for the late twentieth and twenty-first century by such notable figures as the co-designer of the National 9/11 Memorial, Peter Walker, whose five-decade-long career has stressed the dynamics among environmental, social, and economic aspects of a site—is a widely practiced environmental design profession in North America, northern and central Europe, Japan, Korea, and Australia. In China, landscape architecture is rapidly growing from ancient traditions in garden design and site planning. As the world becomes more urban, there are increasing concerns about making cities more livable and protecting both natural and cultural areas. Landscape architecture has a well-established competency in urban design and ecological planning and, as a result, its importance is becoming more broadly recognized.

Traditional and Contemporary Practice

A landscape architecture project begins with a commission or assignment with clear goals, including the scope of the project, proposed uses and users, and the site boundaries and context. The landscape architect may also be involved in site selection. Once the commission begins and the site is selected, the landscape architect then conducts an inventory and analysis of the site. The Scottish-American landscape architect Ian McHarg, author of *Design with Nature* (1969), advocated the use of ecology as a primary guide to structure, including inventories and analyses. An ecological framework enables the landscape architect to understand how physical and biological systems are structured and how they function (Rottle and Yocom 2011). An ecological inventory includes the climate, geology, physiography, ground and surface water hydrology, soils, plants, animals, settlement history, and current land use of the site. The inventory is compiled through maps, diagrams, and written descriptions. It may also involve transects (cross-sectional studies of the site that reveal relationships), among vegetation, drainage, and soils, for example. Inventories are used to conduct suitability analyses that display opportunities and constraints for the proposed uses.

Such site analyses enable the landscape architect to develop various design options. Frequently, these options are used in a formal environmental impact assessment. Many projects are also subject to review by citizens, public agencies, or the commissioning clients. Landscape architects often employ "before and after" drawings and physical models to illustrate their designs' consequences.

The process may conclude in a plan and/or a final design, both of which may require governmental approval for legal or regulatory reasons. A plan then may be implemented through public policy and/or through private actions. A design typically requires detailed construction documents that specify the dimensions and arrangement of its various elements before the project is built.

Traditionally, the maps and drawings were done by hand. In contemporary practice, computer aided design (CAD), computer renderings, geographic information system (GIS) technology, and geodesign techniques are utilized. CAD software systems are used during the design process as well as in design documentation. GIS technologies are essentially computer mapping programs capable of capturing, storing, analyzing, and displaying geographically referenced information.

Innovations

Increasingly, landscape architects are required to demonstrate and measure the outcomes of their designs. The concept of "ecosystem services," the goods and services of direct or indirect benefit to humans that are provided by natural processes involving the interaction of living elements (such as vegetation and soil organisms) and nonliving elements (such as bedrock, water, and air) has proven to be especially helpful in this regard. Examples of ecosystem services include global and local climate regulation, air and water cleansing, water supply and regulation, erosion and sedimentation control, hazard mitigation, pollination, habitat functions, waste decompositions and treatment, human health and well-being benefits, food and renewable non-food products, and cultural benefits.

The Sustainable Sites Initiative (SITES) is an example of the formal application of the ecosystem services concept (Steiner 2011) with the goal of ecological stewardship. Beginning in 2006, SITES was developed through a partnership among the Lady Bird Johnson Wildflower Center of the University of Texas at Austin, the American Society of Landscape Architects, and the US Botanic Garden.

In the SITES system, ecosystem services are linked to specific actions that are considered as prerequisites and credits for SITES certification. The prerequisites and credits affect decisions concerning site selection, predesign assessment and planning, site design, construction, and operations and maintenance to try to minimize aspects of a project that might potentially cause permanent ecological harm, such as the pollution of waterways or destruction of species. Meanwhile, SITES attempts to enhance or maximize any generative or productive project aspects that might produce cultural benefits or enhance the natural environment (such as increased tree cover and recharging of aquifers that supply water). The SITES system establishes uniform, consistent standards, but standards that can adjust to the regional variations of climate, soils, and plants.

Of the sixty-six SITES prerequisites and credits, roughly 60 percent tie quantitative measures of performance to credit achievement, while the other 40 percent are primarily prescriptive in nature; all attempt to tie the attainment of credits with the production of ecosystem services (Windhager et al. 2010).

The credits vary significantly in terms of the performance. Of the thirty-nine credits that set quantitative levels of performance, the bulk remain prescriptive in method, with only seven (21 percent) allowing for open-ended attainment of those performance levels. One example of a high-performance-based credit, "Manage stormwater on site" (credit 3.5), provides a method for comparing regionally adjusted, model runoff-curve numbers for pre- and postdevelopment conditions, and it sets different point values based on preservation or reduction of runoff volumes. This type of credit leaves the landscape architect to determine the ways to achieve performance levels. The landscape architect may choose, for example, to incorporate conventional stormwater approaches (such as detention ponds) or low-impact design approaches (such as rain gardens, rainwater harvesting, or green roofs), so long as the methods used may be shown through modeling to meet the performance goal.

The SITES credits move beyond conservation to the restoration of resources. "Preserve or restore appropriate plant biomass on site" (credit 4.6), for example, intends to ensure regionally appropriate levels of vegetation biomass (referred to as biomass density index or BDI) on site that are sufficient to support ecosystem services. For "greenfield" areas—those never previously developed—postdevelopment vegetation-density levels must at least equal historic predevelopment conditions. For greyfield or brownfield sites that have lost significant levels of vegetation through earlier development, the credit provides a greater array of points based on improvement in the amount of vegetation incorporated into the new site design. Postconstruction vegetation amounts are estimates based on cover type after 110 years of growth and compared to appropriate region-specific vegetation levels based on climate and dominant habitat types. Since landscape architects determine how these biomass-density levels are attained, approaches may include everything from preserving existing areas of high-quality vegetation to creating dense, highly formal gardens, to incorporating green walls and roofs, to some mixture of several approaches.

Outlook

The future of landscape architecture is bright. Ecosystem services provide a restorative lens for landscape architecture practice. The concept of ecosystem services promises to help advance the profession by making its contributions to human and nonhuman health and well-being more explicit. Parks, for instance, have long been viewed as beneficial: as green refuges in the city and as places of recreation. Given a greater common understanding of landscape architecture's aims, parks and other creations of landscape architects can now also be valued for their benefits, such as climate mitigation, improvements in air and water quality, habitat provision, and pollination.

Frederick STEINER
The University of Texas at Austin

See also in the *Berkshire Encyclopedia of Sustainability* Adaptive Resource Management (ARM); Brownfield Redevelopment; Comanagement; Ecosystem Services; Landscape Planning, Large-Scale; Natural Capital; Nutrient and Biogeochemical Cycling; Permaculture; Rain Gardens; Soil Conservation; Stormwater Management; Urban Agriculture; Urban Forestry; Urban Vegetation; Viewshed Protection

FURTHER READING

Carr, Stephen; Francis, Mark; Rivlin, Leanne G.; and Stone, Andrew M. (1992). *Public space*. New York: Cambridge University Press.

Francis, Mark. (2001). A case study method for landscape architecture. *Landscape Journal, 20*(1), 15–29.

Henderson, Harold. (1985). *Prairie speak: The life and art of a forgotten prophet*. Chicago: Chicago Reader Inc.

Hooper, Leonard J. (Ed.). (2007). *Landscape architecture graphic standards*. Hoboken, NJ: John Wiley & Sons.

McHarg, Ian L. (1969). *Design with nature*. Garden City, NY: Natural History Press/Double Day.

Rottle, Nancy, & Yocom, Ken. (2011). *Basics landscape architecture: Ecological design*. West Sussex, UK: AVA.

Steiner, Frederick. (2011). *Design for a vulnerable planet*. Austin: University of Texas Press.

Tishler, William H. (Ed.). (1989). *American landscape architecture*. Washington, DC: National Trust for Historic Preservation.

Windhager, Steven; Steiner, Frederick; Simmons, Mark T.; & Heymann, David. (2010). Toward ecosystem services as a basis for design. *Landscape Journal, 29*(2), 107–123.

Medicinal Plants

Medicinal plants provide traditional medicines and are the origin of several conventional pharmaceuticals used in treating diseases and other afflictions. To sustain this source of medicines, the plant species, plant habitats, and traditional knowledge on the medicinal use of plants need to be protected from overexploitation. In a sustainable medicinal plant system, the plants, people, pharmaceutical companies, patients, and the environment benefit from the continued availability of medicinal plants.

Medicinal plants are a valuable natural resource, the origin of at least 120 chemical substances currently used in conventional pharmaceuticals, and the foundation of the traditional medicine systems used by numerous societies throughout the world. Several familiar drugs illustrate the range of modern pharmaceuticals originally derived from plants, for example, aspirin from willow, used for the treatment of headaches and inflammation; digitoxin from foxglove, used for treatment of heart failure; and vinblastine and vincristine from Madagascar periwinkle, used for treatment of leukemia. More than 60 percent of the world's population depends on traditional plant medicines for primary health care.

The use of plants as medicine has a long history. Knowledge of medicinal plants was passed from generation to generation through stories, rites, and ceremonies before it was recorded in the earliest of writings. Within societies, the shaman, or medicine man or woman, was honored for the knowledge of plant materials that could cure or relieve human suffering. The need for medicines during the deadly plagues of fourteenth century Europe motivated the development of oceangoing ships that could reach the medicinal spices of Asia and ultimately led to the voyages of Christopher Columbus to the North American continent and Vasco da Gama to India in the fifteenth century. The invasion of South America by Spanish conquistadors resulted in the European knowledge that cinchona tree bark, used by the Incas to reduce fever, contained quinine, an effective treatment for malaria.

In modern times the sustainability and diversity needed to protect many medicinal plants is in jeopardy as a result of over-harvest, habitat destruction, biopiracy, and neglect. For example, estimates suggest that 93 percent of the wild plants used in the traditional Ayurvedic medicine system developed in India are threatened with extinction (Khan, Karnat, and Shankar 2005). The loss of a plant species prevents both the discovery of natural chemicals that could serve as medicines, and the ability to use the associated germplasm to improve plants used in the production of pharmaceuticals.

Plants with medicinal value come from a variety of locations and environments. They serve as a worldwide natural warehouse of known and unknown chemical constituents with potential medicinal importance. They also are locally and culturally important. To identify plants that can be used in the treatment of human and animal afflictions, pharmaceutical companies have established bioprospecting programs to identify plants and plant extracts that will be useful in treating human and animal afflictions. Measuring the value of a medicinal plant species or extract in terms of saving a life or curing a disease is difficult, but in a two-year early comparison study (Bishop et al. 1999), taxol from the Pacific yew increased survival for women treated for breast cancer by 19 percent as compared with the standard treatment at that time.

Sustainability

Sustaining the growth of medicinal plants is necessary to provide both the pharmaceutical products and the traditional medicines that protect the health and livelihood of human societies. The World Health Organization has listed 21,000 plant species with known medicinal properties, but only 25 to 30 percent of the 380,800 estimated plant species in the world have been tested (Taylor 2005). Worldwide demand for medicinal plant tissues and extracts for use in pharmaceuticals, foods, dietary supplements, cosmetics, and veterinary products has been increasing at an estimated 3 to 7 percent per year over the past several years. This significant growth can easily lead to unsustainable overcollecting of wild species by individuals and organized groups motivated by the opportunity for financial profit with no apparent concern about the loss of plants.

Global interest and trade in wild medicinal plants can benefit native societies, if done fairly and sustainably, by providing jobs and revenue, but this is not always the case. To be sustainable, the collection methods need to protect the natural environment, ensure continued growth of the plants, and incorporate fair labor and wage controls that enable the people and plants in the ecosystem to flourish. Increased demand and overharvesting practices, however, have sometimes led to overcollection and thievery of plants, particularly in less-developed countries, leaving local people with no plants for traditional medicines and no income. Although the plant species may survive in other locations, decimation of local populations of medicinal plants frequently reduces genetic diversity within the species.

While cultivation can provide for sustainable production, many wild medicinal plants collected in temperate and tropical regions cannot be profitably grown in a field. Field production requires fields and growing seasons suitable for plant growth, as well as plants with uniform seed germination, plant development, harvest maturity, and other desirable agricultural traits that are frequently not present in wild plants and are too expensive to develop. For example, black cohosh seed, which is used to treat symptoms of menopause, needs to be stratified using a warm-cold-warm temperature cycle over three months to induce germination, and then the plant requires three to five years to reach a harvestable size (Greenfield, Davis, and Brayman 2006).

The need for sustainable production for medicinal plants has been recognized. In 1988, health professionals and plant conservation specialists meeting in Chiang Mai, Thailand, affirmed a commitment to primary health care, conservation, and sustainable development, indicating the vital importance of medicinal plants in self-medication and in national health programs (WHO, IUCN, and WWF 1988). To sustain production of medicinal plants, the World Health Organization (WHO 2003) published guidelines for good agricultural and collection practices (GACP) to provide safe and healthful medicinal products, to protect intellectual property rights, and to sustain threatened and endangered species.

No patents are available for plant materials that have been used in traditional medicines for several years because any plant can be patented only if in use for less than one year. In addition, US patent law on plants excludes patenting any plant not being cultivated (Plant Patents 2011). Protection must therefore come through intellectual property rights (WIPO 2011), an important factor in less-developed countries that may have only limited knowledge about the potential markets for plants used in their traditional medicines. If the rights of practitioners to use traditional medicines are not protected, societies that developed the use of plant medicines over several generations could be prevented by law from using the plants in these medicines. Attempts to patent or claim intellectual property rights for commercialization of plants and plant products used in traditional practices is known as biopiracy. An example of biopiracy is the use of neem tree (Azadirachta indica) extracts. While the neem tree is native to the Indian subcontinent and has been grown extensively in India for hundreds of years as a medicine, insecticide, and for other applications, patents were initially granted in Western countries for development of these same types of products from the neem tree. Many of these patents were later revoked by patent offices when presented with evidence on previous uses of neem extracts in India (Hellerer and Jarayaman 2000).

Social and Economic Considerations

Medicinal plants play a significant role in the health, culture, and social and economic livelihoods of many societies. The global market for medicinal plants is also growing due to the increase in human population, acceptance by allopathic (conventional) medical practitioners, and the inadequate provision of conventional Western medicine in developing countries. In the developed world,

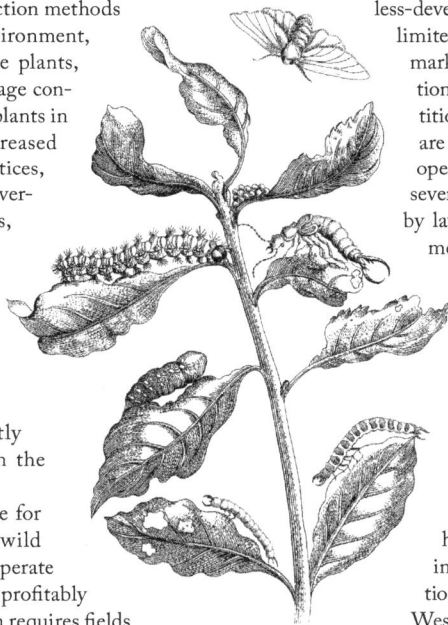

medicinal plants are being used and marketed in health foods, dietary supplements, and cosmeceuticals. Trade in medicinal and aromatic plants, as reported by the twelve leading import and export countries, averaged 320,550 and 368,100 metric tons, respectively, over the period 1991–2003, and frequently the trade is based on products sourced in economically poor countries (Lange 2006).

Many users of medicinal plants wrongly assume that the plant material will always be available. For most endangered medicinal plant species, however, no conservation plans for protecting these plant materials or their habitats is in place. Similarly, knowledge of medicinal plants, acquired over thousands of years by traditional practitioners, is being lost due to the encroachment of conventional Western practices. Relatively few countries have completed inventories of local medicinal species, practitioners, preparations, and medical applications that can be used to support plant conservation and help investigators identify plants and plant extracts useful in the search for medicines. Sustaining medicinal plants and their habitats and the accumulated knowledge about their use will help ensure continued advancements in medical treatments and development of pharmaceuticals.

Lyle E. CRAKER
University of Massachusetts

See also in the *Berkshire Encyclopedia of Sustainability* Drug Production and Trade; Forest Products— Non-Timber; Indigenous and Traditional Resource Management

FURTHER READING

Bishop, James F., et al. (1999). Initial paclitaxel improves outcome compared with CMFP combination chemotherapy as front-line therapy in untreated metastatic breast cancer. *Journal of Clinical Oncology, 17,* 2355–2364.

Greenfield, Jackie; Davis, Jeanine M.; & Brayman, Kari. (2006). *Black Cohosh (Actaea racemosa L.)* (Horticulture Information Leaflet 135). Greensboro: North Carolina State University Cooperative Extension Service.

Hellerer, Ulrike, & Jarayaman, K. S. (2000). Greens persuade Europe to revoke patent on neem tree. *Nature, 405,* 266–267.

Khan, Sarah K.; Karnat, Mohan; & Shankar, Darshan. (2005). India's foundation for the revitalization of local health traditions pioneering in situ conservation strategies for medicinal plants and local cultures. *HerbalGram, 68,* 34–48.

Lange, Dagmar. (2006.). International trade in medicinal and aromatic plants. In R. J. Bogers; L. E. Craker; & D. Lange (Eds.), *Medicinal and aromatic plants* (pp. 155–170). Dordrecht: The Netherlands: Springer.

Native Plant Conservation Campaign. (n.d.). Homepage. Retrieved July 14, 2011, from http://www.plantsocieties.org/

Plant Patents. (2011). FAQ: Nature and duration of plant patents. Retrieved July 28, 2011, from http://www.plant patent.com/faq

Taylor, Leslie. (2005). *The healing power of rainforest herbs.* Garden City Park, NY: Square One Publishers.

World Intellectual Property Organization (WIPO). (2011). Intellectual property and traditional knowledge (Booklet No. 2). Retrieved July 28, 2011 from http://www.wipo.int/freepublications/en/tk/920/wipo_pub_920.pdf

World Health Organization (WHO). (2003). *WHO guidelines on good agricultural and practices (GACP) for medicinal plants.* Geneva: World Health Organization.

World Health Organization (WHO), the World Conservation Union (IUCN) & World Wide Fund for Nature (WWF). (1988). *Guidelines on the conservation of medicinal plants.* Gland, Switzerland: IUCN Publications.

Microbial Ecosystem Processes

Microbial communities can regulate many processes in ecosystems, including decomposition, nutrient cycling, and the degradation of toxic chemicals. These processes all have direct effects on soil and water quality. Microbial processes can be used to promote sustainability, but they also can be sensitive to changes in ecosystems that are caused by human activities. Understanding microbial communities is therefore crucial in managing our world for a sustainable future.

Soil and water contain communities of microbes (microscopic organisms like bacteria and fungi) that make up the majority of the Earth's living matter and account for most of its diversity. These communities are responsible for many important environmental processes, including decomposition of plant and animal material, cycling of nutrients such as nitrogen and phosphorus, and the breakdown of toxic chemicals. Microbial processes contribute significantly to both soil quality (as measured by the structure of the soil and the amount of organic matter and carbon and nutrients present, including nitrogen, phosphorus, and potassium) and water quality (as measured by the amount of nutrients, pollutants, or excess sediments present). Consequently, microbial communities play an important role in how human activities such as growing food, cleaning up oil spills, and addressing climate change will impact natural ecosystems. Understanding microbial processes and using them effectively can greatly improve our ability to manage the Earth's ecosystems in a sustainable fashion.

Microorganism Habitats

Human management of ecosystems directly affects the habitats of microbes for better or for worse. Microorganisms are amazing in that they are adapted to live and function in any habitat on Earth where water, the basis of life, is in liquid form. This includes frozen habitats, where microbes produce sugars to decrease the temperature at which ice can be formed, and deep-sea thermal vents, where extreme pressure allows water to stay in liquid form at temperatures up to 407°C. An important characteristic of microbial habitats is that they contain both *oxic* (containing oxygen) and *anoxic* (containing either low or nearly no oxygen) zones, and a distinct set of microbial processes (termed *aerobic* and *anaerobic*, respectively) occurs in each. Soil and freshwater habitats are described below, as predominant systems of interest to sustainability and as examples of microbial habitats.

Soil

Soil is a living, complex system that provides physical space for microbes to live. The degree to which a soil has structure and how its particles clump together determine how well it is able to withstand erosion. The proportion of mineral particles ranging from very small clay particles (less than 0.002 mm) to large sand particles (0.05–2.0 mm), and organic matter in soil determines what the structure is like. *Organic matter* is a general term for decomposed plants, animals, or microbes; these materials are mainly carbon based, so the carbon can be measured and used to describe how much organic matter a particular soil has. The roots of living plants and filamentous microbes (fungi and filamentous bacteria) also contribute to soil structure by binding soil particles together and strengthening it (bacteria do this by producing complex sugars that they use to "stick" themselves to soil particles). Microbes also contribute significantly to soil formation by carrying out decomposition.

Certain other characteristics of soil (including pH, moisture, and the amount of oxygen present) will affect which microbial processes are most active. These characteristics are determined by climate, position in the landscape (for example, hilltops versus low-lying areas), and local plant communities.

Soil quality (or soil health) is a term often used to describe how degraded an ecosystem is or whether management practices are sustainable. The amount of organic matter in soil, the strength of its structure, and how well it sticks together and retains nutrients determine soil quality. Building a healthy soil creates a positive feedback loop: a soil with lots of organic matter that holds nutrients supports the growth of microbial and plant communities, which create more organic matter and conserve more nutrients and minerals. On the other hand, it's very easy to degrade soil structure (and therefore soil health) through human activities such as agricultural tillage, mining, urban development, and industrial activity.

Water

Microorganisms also are adapted to live in freshwater aquatic and wetland habitats. In lakes, photosynthetic organisms (those that convert carbon dioxide into organic compounds like sugars) are housed mainly in upper layers of the water where light can penetrate. Organisms that depend on organic compounds falling from above live in the deeper layers of the water and in bottom sediments, which can be a nutrient-rich habitat for microbial growth. Life in most lakes is adapted to relatively low nutrient levels, and fertilizer runoff (especially nitrogen and phosphorus) from urban areas or agricultural activities can endanger many forms of life there.

Water in rivers moves at different rates, and most microorganisms either grow attached to rocks or to subsurface sediments. In contrast, wetlands are terrestrial habitats with transient or permanent standing water and plant life that emerges from the water's surface. In wetland habitats, microbes near plant roots in the uppermost layers of water or sediment can consume or transform nutrients rapidly, but these processes also deplete oxygen quickly, so anaerobic processes dominate lower layers. Wetlands are useful in filtering out chemicals or pollutants from urban areas because of this rapid nutrient uptake by microbes and plants, and because wetlands are usually situated in basins or low-lying areas—but at the same time they are also fragile habitats that can be degraded easily by nutrient or sediment runoff.

Microorganism Processes in Ecosystems

Just as microorganisms have adapted to life in diverse habitats, different organisms have adapted so that they can use nearly any material or chemical—from leaf litter (dead plant material) to oil to pesticides—for energy and growth. Microorganisms transform or degrade compounds in the environment for the same reasons that we eat: for energy and to obtain building blocks for growth. Microbes get energy from transforming different compounds based on oxidation-reduction reactions (described in the next section). Large organic compounds are broken down into inorganic building blocks (such as nitrate, ammonia, or phosphate) through *mineralization*.

Oxidation-Reduction Reactions

Oxidation is a process by which the outer electrons from a molecule are stripped away, changing its chemical state. Those electrons are passed to a donor electron acceptor molecule, which becomes *reduced*. Oxidation-reduction reactions yield energy for the microorganism but can also drastically impact whether nutrients or elements stay in the soil or leave, potentially polluting the surrounding environment.

Respiration

Respiration is the use or chemical transformation of any compound for energy production. Aerobic respiration (the use of oxygen as an electron acceptor, resulting in the production of carbon dioxide) is a very efficient process. Where oxygen is available, therefore, microorganisms capable of aerobic respiration will grow quickly and dominate the microbial community. Other chemical transformations used to produce energy, such as nitrification (the conversion of ammonia to nitrate through ammonia oxidation) or methanogenesis (the production of methane gas from carbon dioxide reduction or through acetate fermentation), are much less efficient. Organisms relying on these processes for the energy they need grow more slowly but process large amounts of the compounds used.

Decomposition

Decomposition is the process by which large materials such as dead plants, animals, manure, or microbes are broken down into smaller subunits. Decomposition is nature's way of recycling the nutrients found in living matter back into the soil. Decomposition also can function to store carbon as organic matter or, where the rate of aerobic respiration is high, to deplete supplies of organic matter. Microorganisms create enzymes because large organic molecules, such as cellulose, are too big to be absorbed through the microbial cellular membrane, and many types of microorganisms must produce a variety of enzymes to decompose different organic materials. Some cellular components—such as proteins, cellulose, and starches—are degraded quickly and therefore do not have a long turnover time in the soil before they are degraded and reabsorbed into newly growing microbial cells. Other components such as lignin are not degraded quickly and therefore stay in the soil longer; these help to build up organic matter in soil.

Nitrogen Cycling

Microorganisms are responsible for many transformations of nitrogen that are important to soil and water quality. Nitrogen fixation, mineralization, immobilization, nitrification, and denitrification are all transformations of nitrogen that are important to sustainability, restoration, and soil quality. *Nitrogen fixation* is performed either by free-living bacteria or by bacteria living symbiotically with legumes (a family that includes peas, beans, and clovers). The legumes are able to acquire nitrogen because of their bacterial partner and can improve soil quality or help restore disturbed or low-nutrient soils (such as those on former mining sites). *Nitrogen mineralization* is the process where organic compounds are broken down, releasing inorganic forms that can be used by plants and microbes. *Nitrogen immobilization* is the process by which any soil organism takes up nitrogen so that it's no longer available for other organisms. High rates of immobilization can be positive in that less excess nitrate nitrogen (nitrogen combined in a nitrate ion, as opposed to nitrogen in the form of ammonia, nitrites, etc.) leaks into lakes, rivers, drinking water, or wetlands—but it also can lead to problems with nitrogen limitations for agricultural crops. *Nitrification* is a process linked to pollution that converts nitrogen from a form conserved in the soil (ammonia) to a form that leaches readily from the soil (nitrate) through the energy yielding processes of ammonia and then nitrite oxidation. Often, ammonia fertilizer applied to crops is quickly converted to nitrate. Energy is lost so that crops do not receive the fertilizer they need to grow, and groundwater, lakes, and rivers are polluted by excess nitrogen. Conversely, *denitrification* is a form of anaerobic respiration in which nitrate is reduced into nitrogen gas for energy production. This process also produces nitrous oxide or other nitrogen gases that are considered potent greenhouse gases, but it also can act to remove excess nitrate that otherwise would pollute wetlands and groundwater.

Cycling of Other Elements

Microorganisms act as a middleman in the transformation of many other elements besides carbon and nitrogen. In fact, they can transform nearly any molecule on Earth, and the results can be good or bad for the environment. Acid mine drainage is an example of how the conversion of elements can lead to pollution: when mining or other human activities disrupt mineral rock, microbes transform sulfur (through aerobic respiration) into forms that

are easily released, and water draining from the mine becomes acidic from too much sulfur. Microbes can release environmental toxins such as arsenic and selenium from mineral rock in a similar fashion. Alternatively, microbes contribute to the nutrient levels of soil through mineralization, which transforms sulfur, phosphorus, iron, and other elements from forms that are bound to soil, or breaks down organic molecules into simpler mineral forms that can be taken up by other plants or microbes.

Degradation and Uptake of Toxic Chemicals

Microbes can use toxic chemicals to grow or get energy (through oxidation-reduction reactions) in the same way they use natural elements. These toxins include petrochemicals (such as oil or gasoline), pesticides, metals, and by-products of industrial activity. Sometimes microbial degradation or uptake of these chemicals can remove the contaminant, as with the slow but steady anaerobic degradation of polychlorinated biphenyls (PCBs) that contaminate many soils and aquatic sediments. Microbes can also transform or degrade a toxin only partially, however, resulting in something that is even more toxic or dangerous.

Microorganism Communities and Sustainability

Sustainable management or development has been defined as "the use of land and water to sustain production indefinitely without environmental deterioration" (Lincoln, Boxshall, and Clark 1998). We are managing our environment, for better or for worse, when we decide how to use it. Management can mean policies that are active and promote sustainability (as with agricultural land-management practices and actions that restore degraded landscapes), or it can mean indirect "management" that leads to environmental degradation through global change.

Agricultural Management

Land management practices in agriculture affect microbe communities in a variety of ways. Conventional methods of tilling soil break up the top 20 centimeters of the soil surface and, in the process, severely damage the filamentous cell networks of fungi. Tillage also can lead to organic carbon losses in soil, since it encourages bacterial aerobic respiration by introducing oxygen. These carbon losses can weaken soil structure, leading to erosion and poor-quality soil that doesn't retain nutrients well, resulting in runoff, pollution, and the need to add more chemical fertilizers. Spreading fields with lime changes the pH of soil, affecting which microorganisms are present and how they grow. Using nitrogen fertilizers decreases fungi relative to bacteria, thus altering the normal ratio of processes carried out by each. Nitrogen fertilizer (which is often in the form of ammonium) is then more likely to pollute groundwater, lakes, rivers, and wetlands, as now-abundant bacteria convert it quickly to nitrate. The effects of conventional agricultural cultivation may persist well after it has ended, altering fungal communities, microbial processes, the amount of soil organic carbon, and pH for as much as seventy years.

But agricultural management can become more sustainable. Reducing or eliminating tillage can increase fungal communities, resulting in better soil structure and fungal decomposition processes. Using cover crops (crops that are grown primarily to manage soil quality, weeds, diseases, pests, and water drainage) can prevent soil erosion and add nutrients back to the soil. Operations that grow food can be paired with those that raise animals for milk and food, so that animal waste can be used to add nutrients and organic matter back into the soil. These natural nutrient additions decompose more slowly and thus support long-term improvements in soil quality, unlike mineral fertilizers or repeated tilling. Organic amendments are like an investment bank account that matures over thirty years, while chemical fertilizers are more like a checking account that is drained regularly and needs to be replenished often.

Bioremediation

Using microorganisms to treat waste or pollution is called *bioremediation*, and understanding how microbial processes work is crucial to its success. For instance, certain organisms can be encouraged to grow by providing more or less oxygen, by adding nutrients, or by simply adding the desired organism to a contaminated soil. Sometimes (as with PCBs) just leaving contaminated areas alone and giving organisms time to go about their cleanup work is necessary. Other situations call for balancing different kinds of microbe communities, as when sulfate-oxidizing and sulfate-reducing organisms are paired to remove heavy-metal contamination from soils (a much more effective and cost-efficient process than has been used traditionally). Bioremediation processes don't work in all conditions, but they have huge potential for cleaning up polluted soils.

Human activities can change landscapes greatly even when toxic chemicals aren't involved, and understanding microbial processes can help to rebuild landscapes damaged by mining and other urban activities that lead to a loss of vegetation or excess sediments in wetlands, lakes, or rivers. The ideal path toward better sustainable management, of course, is to be aware of how human activities affect natural ecosystems in the first place and to try to limit the damage that they do.

Impact of Global Changes

Climate change and other global changes could drastically alter which plants can grow in which regions, meaning that agricultural practices will have to be altered and decisions about sustainable land-management policies will become a lot more complex. In addition, microbial communities are probably adapted to local weather in different ways, making it difficult to come up with a one-size-fits-all response to global changes in climate. Climate change also could cause microorganisms to add to the greenhouse gas problem (greenhouse gases are those that absorb radiation, trapping heat in our atmosphere), since their respiration converts stored soil carbon into carbon dioxide, a major greenhouse gas. In places where the climate becomes warmer but stays wet or gets wetter, denitrification and methanogenesis will create more nitrous oxide and methane, respectively, and these are also greenhouse gases. While nitrous oxide and methane are not as abundant as carbon dioxide in the atmosphere, they are much more potent in capturing heat and therefore are important for future changes in our climate.

Another major global change factor is nitrogen deposits from industrial activities. These by-products may act like fertilizers and increase microbial growth and respiration, acidify soil, or pollute rivers and lakes.

The Future

Understanding the many processes mediated by microorganisms can help us use them to make sustainable management decisions. In agriculture, low or no tillage systems with organic inputs to match the nutrients taken away through plant harvesting can improve decomposition processes and microbial contribution to soil structure and stability. Adding nitrogen-fixing microbes or growing legumes (plants with a nitrogen-fixing bacterial partner) to disturbed soil can help to restore it. Microbes can also be used to degrade or remove toxic chemicals that result from industrial waste, mining, and other human activities. Global changes may make microbial communities less able to carry out their usual processes depending on how they adapt. In the future these diverse communities should play a central role in how we think about sustainability and management.

Jessica L. M. GUTKNECHT
Helmholtz Centre for Environmental Research—UFZ

See also in the *Berkshire Encyclopedia of Sustainability* Agricultural Intensification; Agroecology; Biodiversity; Brownfield Redevelopment; Ecosystem Services; Eutrophication; Global Climate Change; Mutualism; Nitrogen Saturation; Nutrient and Biogeochemical Cycling; Plant-Animal Interactions; Soil Conservation; Waste Management

FURTHER READING

Atlas, Ronald M., & Bartha, Richard. (Eds.). (1998). *Microbial ecology: Fundamentals and applications.* Menlo Park, CA: Benjamin/Cummings.

Balser, Teri C.; Gutknecht Jessica L.; & Liang, Chao. (2010). How will climate change impact soil microbial communities? In Geoffrey R. Dixon (Ed.), *Soil microbiology and sustainable crop production* (pp. 373–397). Reading, UK: University of Reading Press.

Balser, Teri C.; Wixon, Devin; Moritz, Lindsey K.; & Lipps, Laura. (2010). The microbiology of natural soils. In Geoffrey R. Dixon (Ed.), *Soil microbiology and sustainable crop production* (pp. 27–57). Reading, UK: University of Reading Press.

Balser, Teri C. (2005). Humification. In Daniel Hillel et al. (Eds.), *Encyclopedia of soils in the environment* (Vol. 2., pp. 195–207). Oxford, UK: Elsevier.

Balser, Teri C; Kinzig, Ann P.; & Firestone, Mary K. (2001). Linking soil microbial communities and ecosystem function. In Ann P. Kinzig, Stephen Pacala, & David Tilman (Eds.), *Linking biodiversity and ecosystem functioning* (pp. 265–358). Princeton, NJ: Princeton University Press.

Banning, N. C.; Grant, C. D.; Jones, D. L.; & Murphy, D. V. (2008). Recovery of soil organic matter, organic matter turnover and nitrogen cycling in a post-mining forest rehabilitation chronosequence. *Soil Biology & Biochemistry, 40,* 2021–2031.

Brussard, Lijbert; de Ruiter, Peter C.; & Brown, George G. (2007). Soil biodiversity for agricultural sustainability. *Agriculture, Ecosystems and Environment, 121,* 233–244.

Burford, E. P.; Fomina, M.; & Gadd, G. M. (2003). Fungal involvement in bioweathering and biotransformation of rocks and minerals. *Mineralogical Magazine, 67,* 1127–1155.

Chander, K., & Brookes, P. C. (1991). Plant inputs of carbon to metal-contaminated soil and effects on the soil microbial biomass. *Soil Biology & Biochemistry, 23,* 1169–1177.

Docherty, Kathryn M., & Gutknecht, Jessica L. M. (2011 [print version forthcoming]). The role of environmental microorganisms in ecosystem responses to global change: Current state of research and future outlooks. *Biogeochemistry.* doi:10.1007/s10533-011-9614-y

Doran, John W., & Zeiss, Michael R. (2000). Soil health and sustainability: Managing the biotic component of soil quality. *Applied Soil Ecology, 15,* 3–11.

Gutknecht, Jessica L. M.; Goodman, Robert M.; & Balser, Teri C. (2006). Linking soil processes and microbial ecology in freshwater wetland ecosystems. *Plant and Soil, 289,* 17–34.

Field, Christopher B., & Raupach, Michael R. (Eds.). (2004). *The global carbon cycle.* Washington, DC: Island Press.

Fry, Stephen C. (2000). The growing plant cell wall: Chemical and metabolic analysis. Caldwell, NJ: Blackburn Press.

Liang, Chao, & Balser, Teri C. (2008). Preferential sequestration of microbial carbon in subsoils of a glacial-landscape toposequence. *Geoderma, 148,* 113–119.

Lincoln, Roger; Boxshall, Geoff; & Clark, Paul. (1998). *A dictionary of ecology, evolution, and systematics* (2nd ed.). Cambridge, UK: Cambridge University Press.

Miller, R. Michael, & Jastrow, J. D. (1992). The application of VA mycorrhizae to ecosystem restoration and reclamation. In Michael Allen (Ed.), *Mycorrhizal functioning: An integrative plant-fungus process* (pp. 438–467). New York: Chapman & Hall.

Nardi, James B. (2007). *Life in the soil: A guide for naturalists and gardeners.* Chicago: University of Chicago Press.

Sylvia, David M.; Hartel, Peter G.; Furhmann, Jeffry J.; & Zuberer, David A. (Eds.). (2004). *Principals and applications of soil microbiology.* Upper Saddle River, NJ: Pearson Education.

Vörös, L., & Szegi, J. (1990). Investigation on the effectiveness of Azotobacter inoculation during the recultivation of mining spoils. *Mikrobiologija, 58,* 642–648.

Mutualism

Mutualism is an interaction between species that benefits both. Such interactions are critical to reproduction and survival and thus to the continued provision of ecosystem services to human populations. Global environmental change threatens mutualisms by altering the timing of natural history events, shifting species' ranges, reducing and fragmenting habitat, and promoting invasion by non-native organisms. More research is needed to accurately predict ecological response to these changes.

Interactions between species influence ecological processes within populations, communities, and ecosystems. Virtually all species on Earth are involved in multiple interspecific interactions at any one time. The best-studied species interactions are competition and predation, relationships that are harmful to one or both of the involved species. Mutualisms, in contrast, are interactions between two species that benefit both. Mutualisms occur in habitats throughout the world (Bronstein 2009) and are crucial to the reproduction and survival of many organisms, as well as to nutrient cycles in ecosystems. Moreover, the services that mutualists provide are increasingly leading environmentalists to consider mutualisms a priority in conservation and restoration.

Ecosystems provide goods and services that support human populations (e.g., food, clean water, energy, protection, and cultural enrichment), without which world economies would "grind to a halt" (Costanza et al. 1997). Some researchers define ecosystem *functions* as processes that facilitate biogeographical flow of energy among ecosystems, while ecosystem *services* are those processes that are beneficial to humans (e.g., Traill et al. 2010). Mutualism plays an important role in ecosystem services

via pollination, seed dispersal, nutrient cycling, and biological control. It is likely that every organism participates directly or indirectly in at least one mutualistic interaction (Bronstein 2009). Since mutualism is involved in virtually every ecosystem service, understanding it and preserving it are of high priority.

Pollination is a classic example of mutualism. Animals pollinate over 87 percent of all flowering plants worldwide (Ollerton, Winfree, and Tarrant 2011). Pollination is essential for agriculture: animal-pollinated crops account for about 35 percent of global food production (Klein et al. 2007). Many plants also depend on animals, including birds and mammals, for seed dispersal. These animals feed on fruits, in the process moving seeds away from parent plants to other habitats, and hence are critical to the persistence of natural and managed vegetation. In another common mutualism, certain plants and insects (such as aphids) use ants for "biological warfare" against their enemies: ants are attracted by food rewards and then aggressively defend their partner against attack. Such interactions can be essential for allowing species to persist, which in turn benefits humans when these are species of economic or aesthetic interest. Conversely, these mutualisms are to our detriment when they involve species (e.g., agricultural pests) that we would prefer to extirpate. Finally, mutualisms between plants and certain microbes promote healthy nutrient and biogeochemical cycling. Mycorrhizae fungi are intimately (symbiotically) associated with plant roots from which plants gain essential nutrients. Approximately 80 percent of terrestrial plants participate in this type of mutualism (van der Heijden 1998). Mycorrhizal fungal diversity is important for the maintenance of plant biodiversity, structure, nutrient capture, and productivity. Similarly, rhizobia are symbiotic nitrogen-fixing bacteria that associate with legumes (including soybeans and peas). These bacteria are responsible for most of the nitrogen that is fixed biologically, as well as for more than a quarter of crop production; thus, they are essential for ecosystem functioning and independence from nitrogen fertilizers. Indeed, mutualistic associations involving microbes are pervasive in nature and are critical to the maintenance of biogeochemical processes, including the nitrogen cycle.

Mutualism is crucial in sustaining coral reefs. Most coral species are dependent on mutualistic associations with specialized symbiotic algae. Environmental stresses, including rising temperatures, cause the expulsion of these algae in a phenomenon known as coral bleaching. While corals may completely recover after mild events, more extreme events can cause 100 percent mortality. Coral reefs provide the structure for highly productive, diverse coastal marine ecosystems that benefit humans by acting as nurseries and habitat for commercial fisheries, as well as recreational habitat. Their destruction through bleaching is expected to become more of a problem for the majority of the world's coral reefs as Earth's climate continues to warm (Donner et al. 2005).

The 2005 Millennium Ecosystem Assessment calculated that human activities have decreased ecosystem services by over 60 percent. Global environmental change negatively affects mutualism in several ways. It alters phenologies (timing of natural history events such as flowering), shifts species' distribution ranges, and reduces or fragments habitat—changes that make it likely that organisms will not be able to find each other and thus to establish the mutualistic associations they require. Furthermore, global change often favors the invasion of species that monopolize or kill mutualists, to their native partners' detriment. Consequences are particularly well documented in seed dispersal and pollination systems. Bee diversity in Britain and the Netherlands has declined since 1980, with a corresponding decline in the plant species reliant on those pollinators (Biesmeijer et al. 2006). Human land use and disturbance are significant threats to wild, native pollinators (Winfree et al. 2009), and thus to plant communities worldwide. Biologists have reported a few cases of phenological mismatches among plants and their pollinators as a consequence of climate warming (Hegland et al. 2009), although there are also cases known in which partners shift their own phenology to the same degree, leaving their associations intact.

Biologists have recently urged that species interactions need to be taken into account if we wish to accurately predict ecological response to global change (Kiers et al. 2010). Little is yet known about the effects of global change on biotic interactions, especially mutualism. Plant-pollinator systems have received more attention than any other mutualistic interaction; close attention to the size of intact habitat and ecosystem-specific thresholds has been shown to be essential for maintenance of wild pollinators

and pollinator-dependent plants in agro-ecosystems (Keitt 2009). Continued accumulation and analysis of data for other biotic interactions will lead to understanding of what is needed to protect and sustain other ecosystems as well.

Ginny M. FITZPATRICK and Judith L. BRONSTEIN
The University of Arizona

See also in the *Berkshire Encyclopedia of Sustainability* Agroecology; Biodiversity; Charismatic Megafauna; Community Ecology; Disturbance; Global Climate Change; Food Webs; Habitat Fragmentation; Indicator Species; Keystone Species; Invasive Species; Microbial Ecosystem Processes; Nutrient and Biogeochemical Cycling; Outbreak Species; Plant-Animal Interactions; Resilience

FURTHER READING

Biesmeijer, Jacobus C., et al. (2006). Parallel declines in pollinators and insect-pollinated plants in Britain and the Netherlands. *Science, 313*(5785), 351–354.

Bronstein, Judith L. (2009). Mutualism. In Simon A. Levin (Ed.), *The Princeton guide to ecology* (p. II.11). Princeton, NJ: Princeton University Press.

Costanza, Robert, et al. (1997). The value of the world's ecosystem services and natural capital. *Nature, 387*(6630), 253–260.

Donner, Simon D.; Skirving, William J.; Little, Christopher M.; Oppenheimer, Michael; & Hoegh-Guldberg, Ove. (2005). Global assessment of coral bleaching and required rates of adaptation under climate change. *Global Change Biology, 11*(12), 2251–2265.

Hegland, Stein Joar; Nielsen, Anders; Lázaro, Amparo; Bjerknes, Anne-Line; & Totland, Ørjan. (2009). How does climate warming affect plant-pollinator interactions? *Ecology Letters, 12*(2), 184–195.

Kearns, Carol A.; Inouye, David W.; & Waser, Nickolas M. (1998). Endangered mutualisms: The conservation of plant-pollinator interactions. *Annual Review of Ecology and Systematics, 29*, 83–112.

Keitt, Timothy H. (2009). Habitat conversion, extinction thresholds, and pollination services in agroecosystems. *Ecological Applications, 19*(6), 1561–1573.

Kiers, E. Toby; Palmer, Todd M.; Ives, Anthony R.; Bruno, John F.; & Bronstein, Judith L. (2010). Mutualisms in a changing world: An evolutionary perspective. *Ecology Letters, 13*(12), 1459–1474.

Klein, Alexandra-Maria, et al. (2007). Importance of pollinators in changing landscapes for world crops. *Proceedings of the Royal Society B: Biological Sciences, 274*(1608), 303–313.

Kremen, Claire, et al. (2007). Pollination and other ecosystem services produced by mobile organisms: A conceptual framework for the effects of land-use change. *Ecology Letters, 10*(4), 299–314.

Millennium Ecosystem Assessment. (2005). Ecosystems and human well-being: Synthesis. Washington, DC: Island Press.

Ollerton, Jeff; Winfree, Rachel; & Tarrant, Sam. (2011). How many flowering plants are pollinated by animals? *Oikos, 120*(3), 321–326.

Traill, Lochran W.; Lim, Matthew L. M.; Sodhi, Navjot S.; & Bradshaw, Corey J. A. (2010). Mechanisms driving change: Altered species interactions and ecosystem function through global warming. *Journal of Animal Ecology, 79*(5), 937–947.

van der Heijden, Marcel G. A., et al. (1998). Mycorrhizal fungal diversity determines plant biodiversity, ecosystem variability and productivity. *Nature, 396*(6706), 69–72.

Winfree, Rachel; Aguilar, Ramiro; Vázquez, Diego P.; LeBuhn, Gretchen; & Aizen, Marcelo A. (2009). A meta-analysis of bees' responses to anthropogenic disturbance. *Ecology, 90*(8), 2068–2076.

Natural Capital

Natural capital is an economic construct that describes the natural world, its ecosystems, and their value to society. How people value the natural world determines how businesses and societies both conserve and deplete it. Economists who think about natural capital as an irreplaceable resource and those who believe that it is like any other input into an economy have very different ideas about how society should treat the natural world.

Economists use the concept of natural capital to explain the contribution that the natural world's resources make to human economies. Different schools of economic thought have a number of different ways to approach the topic, and these approaches have different consequences for sustainable development.

Conceptual History

In the eighteenth and nineteenth centuries, economists identified the factors of production (that is, the resources that go into producing goods and services), as capital, labor, and land. Capital was defined as an input that is not consumed in the manufacture of a product (Smith 1776) or, alternatively, as something human made that contributed to production (Böhm-Bawerk 1891), for example, machinery. Land, which included all natural resources, was treated as distinct from capital because it was a gift of nature and because humans could not affect its supply. In the twentieth century, economists redefined capital as any asset that produces a stream of income over time (Fisher 1906). By this definition, land is lumped together with other capital, and the factors of production are reduced to two: capital and labor. After the redefinition, natural resources were increasingly ignored as a factor of production to the point where an eminent

economist could suggest that "the world can, in effect, get along without natural resources" (Solow 1974). In the 1970s, however, growing evidence of the limitations on natural resources and environmental problems made worse by accelerating economic growth led many economists to call for explicitly recognizing natural capital—defined as a stock that yields a flow of natural services and tangible natural resources—as a distinct and essential factor of production.

The first explicit reference to natural capital appears in *Small Is Beautiful* (Schumacher 1973). In this book, the British economist E. F. Schumacher argued that irreplaceable natural capital stocks make up the larger part of all capital, and that modern economists erroneously treat their depletion as income. Schumacher identified two types of natural capital. The first was fossil fuels, which were rapidly being exhausted. The second was the ability of natural systems to regenerate themselves, threatened by novel chemicals against which nature had no defenses. Although they did not use the specific phrase *natural capital*, other researchers, including Herman Daly (1973, 1977) and Nicholas Georgescu-Roegen (1971) were simultaneously stressing that the goods and services provided by nature are essential, nonsubstitutable factors of production, and that the finite supply of these resources limits continued economic growth. Furthermore, both Daly and Georgescu-Roegen carefully distinguished between the two types of natural capital discussed by Schumacher. Fossil fuels, along with all other raw materials from nature (both renewable and nonrenewable), are identified as stock-flow resources, which are consumed and therefore depleted in the act of production. Humans can decide how quickly to deplete such resources. In contrast, the ability of ecosystems to reproduce themselves, along with other services provided by ecosystems, is a fund-service resource. A fund is not consumed in the act of producing a service. For example, when a forest helps regulate water flow, processes waste, provides shelter for other species, or produces the seeds required for renewal, it is not consumed in the process. Nature's fund services are generated from a particular configuration of its stock-flow components. Just as a car factory is a particular configuration of metal, glass, and concrete, a forest is a particular configuration of plants, animals, water, and soil. Funds provide services at a given rate over time. The term *natural capital* generally refers both to stock flows and to fund services.

The concept of natural capital caught on fairly quickly, particularly in the field of ecological economics, whose theoretical foundations stressed the dependence of the economic system on the planet's finite supply of natural resources and the invaluable services they generate. The economists David Pearce (1988) and Herman Daly (1990) argued that sustainable development required a constant natural capital stock. Daly listed specific rules for maintaining a constant stock: extraction of renewable resources could not exceed regeneration rates, extraction of nonrenewable resources could not exceed the rate at which renewable substitutes were developed, and waste emissions could not exceed the ecosystem's absorption capacity. The concept of natural capital suited ecological economic theory so well that the proceedings of the second international Ecological Economics Conference were published as the book *Investing in Natural Capital* (Jansson et al. 1994). The first section of this book focuses on maintaining and investing in natural capital, the second focuses on methods and research topics, and the third on policy implications and applications. These three topics parallel and anticipate the way researchers later developed the ideas around natural capital.

Much of the early research on natural capital focused on the economic value of ecosystem services. In May 1997, the journal *Nature* published a paper that integrated these studies into a single global assessment of natural capital. That paper, "The Value of the World's Ecosystem Services and Natural Capital" (Costanza et al. 1997), has become one of the most cited in the environmental sciences.

At the same time, various nations attempted to integrate natural capital into their national accounts (Ahmad, El Serafy, and Lutz 1989), leading the United Nations to propose a System of Environmental and Economic Accounts (SEEA) (United Nations 1993), eventually implemented in 2003. The researchers William Rees (from Canada) and Mathis Wackernagel (from Switzerland) introduced the concept of the ecological footprint as a biophysical measure of humanity's demands on natural capital (Rees and Wackernagel 1994). A growing number of countries, regions, and businesses around the world adopted the ecological footprint as a measure of sustainability.

The book *Natural Capitalism* (Hawken, Lovins, and Lovins 1999) and the nonprofit organization The Natural Step both played a critical role in popularizing the concept of natural capital outside of academia, particularly in the business world, by laying roadmaps for a society that obeys Herman Daly's specific rules for sustaining natural capital. The triple bottom line is one increasingly popular approach to business that accounts for natural capital, human capital, and financial capital (Elkington 1997). National and international policies designed to protect and restore natural capital including cap and trade regulations for pollutants and fisheries and payments for ecosystem services are now multibillion-dollar global markets (Farley et al. 2010).

Strong and Weak Sustainability

Natural capital, however, remains poorly defined and subject to controversy. One ongoing debate is whether or not natural capital is, in fact, irreplaceable. If it is true that some natural capital is essential and has no substitutes, then that capital must be preserved and so cannot be lumped together with other forms of capital. This belief, generally shared by ecological economists and known as strong sustainability, led to the emergence of the concept of natural capital in the early 1970s. Other economists argue that human-made capital can substitute for natural capital and that as long as the value of both types of capital together is not declining, sustainability is achieved. According to this model, clearcutting the Amazon could be viewed as sustainable as long as future generations were left with an equal value of roads and buildings. This approach is called weak sustainability. David Pearce was instrumental in developing the concept of weak sustainability (Pearce and Atkinson 1993), although he had initially stressed the irreplaceable nature of natural capital. Many scholars now use the phrase *critical natural capital* for natural capital that is essential to human welfare and has no substitutes (Ekins et al. 2003).

A related debate concerns the labels *natural capital* and *ecosystem services*, which, some argue, imply the treatment of nature as a commodity, or market good, and hence weak sustainability. Many economists do believe that natural capital should be treated as just another commodity and incorporated into the market model. Others, however, interpret natural capital as a metaphor that calls attention to the productive capacity of ecosystems and the need to invest in their protection and restoration. If natural capital is defined as an asset that produces a flow of income over time, then the term implies that we must live only off the flow of income without depleting the capital stock. According to this way of thinking, the metaphor does not imply that natural capital can be bought and sold like any other asset.

Valuing ecosystem services goes a step further than the natural capital metaphor to suggest that such services have a monetary exchange value and are neither essential nor nonsubstitutable. Advocates of weak sustainability typically believe that markets will lead to the optimal provision of ecosystem services if the services are correctly priced. Even many advocates of strong sustainability argue that valuation calls attention to the importance of natural capital and that failure to value ecosystem services assigns an implicit value of zero to them. They point out that food is also essential and nonsubstitutable, yet it is nonetheless valued in monetary terms.

Critics of monetary valuation argue that it is based on preferences weighted by purchasing power, which gives no voice to the poor and prioritizes Western over indigenous values. Furthermore, preferences are often based on incomplete or inaccurate knowledge, because people rarely understand precisely how ecosystems generate services or how human activities will affect them. The dominant critique is that valuation implies weak sustainability. In fact, many conventional economists focused on dollar values have explicitly stated that global climate change is relatively unimportant because it primarily affects agriculture, which constitutes a negligible share of gross domestic product, or GDP (Schelling 2007). Most ecological economists argue that society should impose the specific quantitative rules for sustainable use of natural capital that Herman Daly suggested, and then let prices adjust to these ecological constraints (Farley 2008). They also suggest that because natural capital is part of a shared inheritance, scientific and democratic principles, rather than the market, should be used to value it.

Another controversial application of the concept of natural capital is payments for ecosystem services (PES). Those who favor the integration of natural capital into the market tend to favor PES systems in which private sector beneficiaries of ecosystem services pay landowners for land uses that provide specific services. Such payments are often for a single service and do not take into account other services provided by the system. Critics of PES typically argue that ecosystem services are public goods and cannot be forced into the market model. Protecting and restoring natural capital does, however, impose real costs on society, which somebody must pay. Many ecological economists believe that investments in natural capital should be a cooperative endeavor, in which the wealthiest nations and regions shoulder the financial cost of restoration and conservation wherever it is needed (Farley and Costanza 2010).

Outlook for the Future

The importance of the natural capital concept continues to grow, as measured by a steadily increasing use of the term in the scientific literature. Human society must recognize that we, like all other species, depend on the flow of goods and services provided by nature. In the past, humans have treated natural capital as if there were enough to meet all human and ecological needs for all time, with no trade-offs involved and hence no need to ration access. The market system is very effective at allocating natural capital toward market products, but it fails to take account of natural capital's growing scarcity. As a result, societies are now depleting natural capital faster than it can regenerate and returning waste to the environment faster than it can be absorbed. Depletion of this natural capital will diminish not only nature's capacity to regenerate itself but also the raw materials needed for all economic production and the flow of ecosystem services essential to human wellbeing. Future generations are left dependent on dwindling resources. The concept of strong sustainability makes it obvious that people must learn to live on the interest from natural capital, the annual flow of benefits, without depleting the stock.

Natural capital is inherently different from other forms of capital—produced capital and natural capital are ultimately complements, not substitutes for one another. It is not enough to simply put a price on natural capital and force it into competitive market boundaries. Instead, economic institutions must adapt to the fact that natural capital is irreplaceable and generates a flow of public good services best protected by cooperative efforts. Society faces a new allocation challenge: how much natural capital should be converted to economic production, and how much should be conserved for the provision of ecosystem

services? Both uses of natural capital are essential and have no substitutes. The concept of natural capital, correctly applied, can help society make these choices.

Joshua FARLEY
University of Vermont

See also in the *Berkshire Encyclopedia of Sustainability* Agricultural Intensification; Agroecology; Ecosystem Services; Human Ecology; Irrigation; Landscape Planning, Large-Scale; Marine Protected Areas (MPAs); Ocean Resource Management; Permaculture; Reforestation; Soil Conservation; Viewshed Protection; Water Resource Management, Integrated (IWRM); Wilderness Areas

FURTHER READING

Ahmad, Yusuf J.; El Serafy, Salah; & Lutz, Ernest. (Eds.). (1989). *Environmental accounting for sustainable development.* Washington, DC: The World Bank.

Böhm-Bawerk, Eugen von. (1891). *The positive theory of capital.* London: Macmillan.

Costanza, Robert, et al. (1997). The value of the world's ecosystem services and natural capital. *Nature, 387,* 253–260.

Daly, Herman E. (1977). *Steady-state economics: The economics of bio-physical equilibrium and moral growth.* San Francisco: W. H. Freeman.

Daly, Herman E. (1990). Towards some operational principles for sustainable development. *Ecological Economics, 2*(1), 1–6.

Daly, Herman E. (Ed.). (1973). *Toward a steady-state economy.* San Francisco: W. H. Freeman.

Ekins, Paul; Simon, Sandrine; Deutsch, Lisa; Folke, Carl; & De Groot, Rudolf. (2003). A framework for the practical application of the concepts of critical natural capital and strong sustainability. *Ecological Economics, 44*(2–3), 165–185.

Elkington, John. (1997). *Cannibals with forks: The triple bottom line of 21st century business.* Oxford, UK: Capstone Publishing.

Farley, Joshua. (2008). The role of prices in conserving critical natural capital. *Conservation Biology, 22*(6), 1399–1408.

Farley, Joshua; Aquino, André; Daniels, Amy; Moulaert, Azur; Lee, Dan; & Krause, Abby. (2010). Global mechanisms for sustaining and enhancing PES schemes. *Ecological Economics, 69*(11), 2075–2084.

Farley, Joshua, & Costanza, Robert. (2010). Payments for ecosystem services: From local to global. *Ecological Economics, 69*(11), 2060–2068.

Fisher, Irving. (1906). *The nature of capital and income.* New York: Macmillan.

Georgescu-Roegen, Nicholas. (1971). *The entropy law and the economic process.* Cambridge, MA: Harvard University Press.

Hawken, Paul; Lovins, Amory; & Lovins, L. Hunter. (1999). *Natural capitalism: Creating the next industrial revolution.* Boston: Little, Brown.

Jansson, AnnMari; Hammer, Monica; Folke, Carl; & Costanza, Robert. (Eds.). (1994). *Investing in natural capital: The ecological economics approach to sustainability.* Covelo, CA: Island Press.

Pearce, David. (1988). Economics, equity and sustainable development. *Futures, 20*(6), 598–605.

Pearce, David W., & Atkinson, Giles D. (1993). Capital theory and the measurement of sustainable development: An indicator of "weak" sustainability. *Ecological Economics, 8*(2), 103–108.

Rees, William E., & Wackernagel, Mathis. (1994). Ecological footprints and appropriated carrying capacity: Measuring the natural capital requirements of the human economy. In AnnMari Jansson, Monica Hammer, Carl Folke & Robert Costanza (Eds.), *Investing in natural capital: The ecological economics approach to sustainability* (pp. 362–390). Covelo, CA: Island Press.

Schelling, T. C. (2007). Greenhouse effect. In D. R. Henderson (Ed.), *The concise encyclopedia of economics.* Indianapolis, IN: Liberty Fund.

Schumacher, Ernst Friedrich. (1973). *Small is beautiful: Economics as if people mattered.* New York: Harper and Row.

Smith, Adam. (1776/1996). *An inquiry into the nature and causes of the wealth of nations.* Oxford, UK: Clarendon Press.

Solow, Robert M. (1974). The economics of resources or the resources of economics. *The American Economic Review, 64*(2), 1–14.

United Nations. (1993). *Handbook of national accounting: Integrated environment and economic accounting* (Series F, No. 61). New York: United Nations.

Nutrient and Biogeochemical Cycling

Understanding nutrient and biogeochemical cycling (the movement of elements through ecosystem components) can provide relatively unambiguous and simple criteria for evaluating one aspect of sustainability. Based on a "mass balance" approach (which looks at the balance of ecosystem inputs and outputs), descriptions of elemental cycles over human time frames can determine whether management practices will result in losses in nutrient stocks (the total ecosystem inventory of nutrients in various forms) that will ultimately degrade ecosystem function.

The nutrient aspects of sustainability in land management (e.g., forestry, agriculture) perhaps provide the most unambiguous criteria for developing management plans consistent with sustainability principles. The contrasts between "strong" and "weak" sustainability—the proponents of the former saying that the Earth's natural capital (the stock of natural ecosystems that yields a flow of valuable ecosystem goods or services) is irreplaceable, and proponents of the latter saying that economic growth can be accommodated by substituting technology and other resources for depleted natural capital—are less controversial in this context as nutrients are generally not substitutable. For example, if nitrogen is in short supply, extra phosphorus will not compensate for that limitation. The *law of the minimum* posited by the German chemist Justus von Liebig (1803–1873) provides the framework for this concept of nonsubstitution: growth will continue until the nutrient in the shortest supply becomes limiting; resupply of that nutrient will renew growth. Thus, understanding the basic dynamics of nutrients on human time scales provides a solid foundation for planning and management of ecosystems that would be consistent with principles of sustainability. More specifically, management of resource extraction to insure that nutrient stocks

are not depleted and remain consistently available for continued net primary production (total plant growth via photosynthesis minus respiration of the plants) would be a key criteria of sustainable management. Because net primary productivity fuels the rest of the ecosystem (including primary, secondary, tertiary, etc., consumers and organisms that consume dead material and decompose organic matter), maintaining current levels of productivity sustains the entire ecosystem over time. Using a nutrient criterion for sustainability thus complements other sustainable management issues such as erosion, soil compaction, pollutant contamination, and loss of biodiversity.

Thinking like an Ecosystem

Prior to the advent of the ecosystem concept, concerns about forestry and agriculture focused more narrowly on issues such as species management and soil fertility. Foresters worried that after a tree harvest, their favored species would not regenerate to produce another crop of desirable trees to cut down the next time around. Agriculturalists worried that soil fertility would decline, and they would either have to find new land to till or purchase expensive fertilizers to apply to the land. The soil test was one solution to this concern. If the concentration of nutrients in the soil sample declined, then the remedy was to add the limiting nutrient (Liebig's law again).

The foundation for understanding forests and agricultural fields in a different way—through knowledge of nutrient and biogeochemical cycling—began with the adoption of the ecosystem concept as a way of looking at forests, fields, and other areas. Articulated in 1935 by the British ecologist Arthur Tansley and experimentally employed to quantify ecosystem function in the 1960s (e.g., see Bormann and Likens 1967), the ecosystem concept provides the framework for describing nutrient stocks both conceptually and quantitatively. This concept focused on a bounded place as an interacting system of living and physical components (biotic and abiotic) that could be characterized by stocks and flows of matter and energy. For example, while the agriculturalist may have been happy when the concentration of a limiting nutrient was maintained at an optimal level, the actual stock of nutrients was declining as the depth of the soil was diminishing. This is analogous to evaluating the health of your bank account by the willingness of the bank teller to let you withdraw one hundred dollars a week. Sustainability requires you to keep track of the account balance.

By contrast, in using an ecosystem approach, places like forests and crop fields were studied by carefully establishing ecosystem boundaries and components, and then characterizing major inputs and outputs of substances like nutrients. In retrospect, this seemingly simple research perspective of defining a bounded ecosystem and considering inputs and outputs to be like deposits and withdrawals from the bank provided important new insights into the functioning of both natural and human-managed ecosystems. For example, early work by Harold Hemond (1980) on nutrient cycling in Thoreau's Bog in Concord, Massachusetts, determined that it was ombrotrophic (rain fed), which accounted for the unusual plants, animals, and nutrient-retention survival strategies.

In addition, ecosystem analysis provided an integrative perspective on how living organisms and their abiotic components worked together. An ecosystem is a complex assemblage of hundreds and perhaps thousands of species of plants, animals, and microbes, along with the physical complexity of soil horizons, parent material, and ever-changing atmospheric conditions. The ecosystem approach simplified this complexity by focusing on a relatively small set of ecosystem components. This new understanding would eventually challenge existing management practices as described in sections below.

New Insights

New insights gained from studying nutrient cycling in ecosystems formed the basis for questioning the sustainability of human practices on the landscape. Not long after the development of techniques for ecosystem analysis in the late 1960s, F. Herbert Bormann and collaborators applied the ecosystem framework to the practice of clear-cutting. In one of many experiments they found manyfold increases in nutrient losses following cutting of trees. This finding generated a great deal of controversy and subsequent research, as well as new policies governing forest management. Clear-cutting could destabilize the biotic control of the cycling and retention of those nutrients, causing a loss of nutrients not only in the removal of the harvested wood but also in the leaching of nutrients into

stream water. Another concern reported by the National Academy of Sciences in 1980 involved the conversion of moist tropical forest areas. In the previous decade, the ecologist Nellie Stark (1971a; 1971b) determined that the stock of nutrients in living biomass in Amazonian forests was very large in comparison with the stock of nutrients in the heavily weathered soils. This implied that removal of living biomass from the ecosystem might seriously jeopardize the ability of the forest to regenerate itself, given its reduced stock of nutrients. This was of particular concern with the conversion of primary forest, with their huge pools of nutrients in biomass, to secondary forest that then needed to re-accumulate nutrients from yet unquantified sources of inputs. In many areas, tropical deforestation may be unsustainable from a nutrient perspective.

Studies of nutrient dynamics in the context of management emphasized the need for research into the basic ecosystem dynamics of a whole suite of elements, including both nutrients and environmental contaminants. This new approach to understanding ecosystems has been subsequently applied to understanding acid rain, nitrogen saturation (over-fertilization), mercury pollution, and the biggest hurdle of the twenty-first century, global climate change.

Sustainability: Nutrient and Biogeochemical Cycling

As suggested above, nutrient cycling criteria may provide some of the most rigorous and unambiguous criteria for the evaluation of sustainability. Considering specific nutrients one at a time, we can use techniques of ecosystem analysis to achieve our best understanding of ecosystem function for that element. The joint consideration of many nutrient elements can be approached in the context of the previously mentioned Liebig's law of the minimum. The measurement of the stock of this limiting factor can inform managers whether current practices are sustainable over the long term.

In the almost infinite diversity of natural and human-created ecosystems, the nutrient factors limiting primary productivity may vary considerably. One simplification is to consider two categories of nutrients: macronutrients (nutrients needed by life in large quantities, e.g., nitrogen, phosphorus, potassium, calcium, magnesium, sulfur) and micronutrients (nutrients needed in only minute quantities, e.g., boron, copper, iron, chlorine, manganese, molybdenum, zinc). Choosing a few to be illustrative of the application of nutrient and biogeochemical cycling to sustainability can provide the foundation for developing appropriate sustainability criteria.

In addition, as mentioned above, contaminants cycling is also a consideration for thinking about sustainable management. For example, natural wetland ecosystems are increasingly valued for their pollutant "filtering" capacity, and constructed wetlands are often designed specifically for these functions. Understanding cycling in these contexts is essential for managing contaminants over the long term in relation to sustainability principles.

The terms *nutrient cycling* and *biogeochemical cycling* need some further discussion as they are both used in the title of this section. As used here, the distinctions in the terms lie in how they may be applied to nutrients versus contaminants. *Biogeochemical cycling* is the broader of the terms as it applies to any element moving through the biosphere and lithosphere, whether it is a nutrient of life or not. Thus the biogeochemical cycling of a contaminant like mercury (Hg) is not appropriately considered nutrient cycling as there is currently no known need of mercury in animal or plant physiology. *Nutrient cycles* on the other hand have biological, geological, and chemical components to their cycles, and thus are all properly biogeochemical cycles. Therefore, when discussing nutrient elements, context and semantic preference will govern the use of either term.

Cycling and Time

Most advocates of sustainability work in the context of human time scales. Even the long view of some native peoples (some who are said to plan in the span of "seven generations") is framed within a time scale relevant to social processes and not geologic processes. Geologic processes figure prominently in the cycles of most nutrient elements however. Thus understanding biogeochemical cycles over large spans of time is critical to developing an understanding of sustainability and cycling.

For example, at human time scales, calcium cycling is not a cycle at all. Calcium weathers out of rocks, goes into solution, gets carried by ground-, subsurface, and surface water to the ocean, and then precipitates or settles on the seafloor through biotic and chemical processes. While this march to the ocean can be interrupted by ecological processes and recycled to the terrestrial system to then restart the journey to the sea, the vast majority of calcium atoms take a one-way trip to the sea. This depletion of calcium, like the mining of fossil fuels, degrades the pool of calcium with each successive millennium, weathering towering alps to gentle green mountains to pleasant hills and valleys. Also known as a sedimentary cycle to geologists, this calcium story is very much a cycle as the calcium-rich sediments accumulate to great depths over hundreds of millions of years, form rocks, and may eventually get uplifted by tectonic forces and reemerge as towering mountains that will weather the forces of physical erosion and acid dissolution, and thus start the process all over again. Thus sustainability principles applied in the context of nutrient and biogeochemical "cycles" must consider both the maintenance of elemental stocks and their dynamics (e.g., weathering).

The variability of nutrient dynamics in many developing ecosystems also requires consideration in the context of sustainability. For example, if we harvest trees on a continuous cycle in a way that does not deplete the pool of nutrients, thereby sustaining subsequent productivity for the next forest cycle, we might all agree that at least from a nutrient perspective, the practice is sustainable. But even in the relatively rapid span of human lives, natural ecological processes change the status of nutrient pools. For example, Bernard Bormann (the son of F. Herbert Bormann) and collaborators (1998) documented that primary succession of forest over sand can rapidly build biomass and available nutrient pools. Thus the "normal" dynamics of nutrient stocks, including available soil nutrients, would be increasing. Would sustainability criteria focus on increasing nutrient stocks or just maintaining a constant level? In another example, natural fires deplete ecosystem nitrogen (through conversion to nitrogen oxides that are lost to the atmosphere), lowering total nitrogen stocks below what would be present without fire. Right after fire, however, mineralized nutrients boost recovery through short-term fertilization. Thus managing fire frequency sustains different levels of productivity. What is the norm that we would want to manage sustainably? In both examples above, management may sustain stocks of nutrients in the ecosystem at static levels, where natural ecosystem development may not sustain constant stocks. Thus there are specific cases where principles of sustainability may require a more dynamic interpretation of nutrient "cycling."

Sedimentary Cycles

In a geologic time frame, cycles have been categorized as sedimentary cycles and gaseous cycles, with some cycles having important elements of both. Each has some important distinctions in the context of sustainability. The calcium cycle, introduced above, is a good example of a sedimentary cycle. Elements categorized as having a sedimentary cycle have a very long cycling time associated with weathering from rock substrates, transport in surface waters, sedimentation in ocean waters, rock formation, uplift, and back to weathering.

For a terrestrial system, the implication of this long-term biogeochemical cycle is that nutrients need not be conserved within the ecosystem. Outputs will exceed inputs as the storage pool of nutrients in rocks is slowly depleted. Limestone weathering provides a simplified model of the sedimentary cycle starting with the following carbonate weathering equation:

$$CaCO_3 + H^+ + HCO_3^- \longrightarrow Ca^{2+} + 2HCO_3^-$$

Calcium (in ionic form: Ca^{2+}) participates in the internal or intrasystem cycle within the ecosystem, where calcium drives growth of plants and feeds the nutritional needs of the many animals, fungi, and microbes that make up the ecosystem. But the net export puts these positively charged calcium cations into the surface waters of the world to then travel to the ocean. During this journey to the ocean, calcium continues to feed aquatic plant growth and all the organisms that depend on this productivity. Reaching the ocean, calcium is an important nutrient for many organisms, but to feed the long-term biogeochemical cycle, there are key organisms, such as diatoms, that use calcium as structural components. Upon their death, these calcium-rich structures settle to the bottom of the sea and begin the rock-formation process. Millions of years later, with an uplift event, these calcium atoms return to feed terrestrial organisms through the weathering process.

Thus, sustaining ecosystem productivity requires not the conservation of calcium, but the sustainable stewarding of the element on its way to the sea. Where humans seek to remove products of biology such as trees or crops from the ecosystem, the normal weathering rate might be a rough measure of how much can be removed without diminishing the stock of available calcium. This recognition of the role of weathering in nutrient cycling was first introduced by Nellie Stark (1978) with the proposal of a "biological life of soils." Working in both young, recently glaciated forest soils and heavily weathered soils in the tropics, Stark described the natural evolution of sedimentary cycles and proposed that removal of nutrients through (a) crop or tree harvest and (b) land management practices that could increase or decrease weathering rates were important components of balancing human use and natural processes.

The components of this balance involve weathering rates, other natural (e.g., nutrients carried in precipitation) and anthropogenic inputs (e.g., particulate pollution), harvest removals, and aqueous outputs (e.g., calcium leaving the ecosystem in surface and ground water). This balance should be evaluated over the cycle of harvests (e.g., rotation rate of tree harvests) to evaluate sustainability. In between harvests, the balance also includes the accumulating storage of calcium in living and dead biomass. But if we assume no net change in biomass over the course of a harvest rotation, then the most important change in storage will be the available pool of calcium. A simplified mass balance equation can assist in conceptualizing the role of management on nutrient balance:

weathering + inputs + depletion of nutrient stock
= harvest + outputs

(Inputs include natural phenomenon like precipitation and anthropogenic sources such as particulate pollution; outputs include losses to stream water and groundwater.)

Clearly, excessive harvests or human impacts that increase outputs of particulate or dissolved nutrients will deplete the stock of available calcium. If calcium scarcity, resulting from a decline in the available stock, reduces primary productivity, then the management practices could be considered to be unsustainable.

An interesting dimension of this mass balance model of ecosystem dynamics is that the stock of available calcium is not tied to harvest levels. As long as harvests and aqueous losses do not exceed weathering and other inputs, the ecosystem remains in nutrient balance. Thus this balance could be achieved under conditions of either low nutrient availability *or* high nutrient availability. Clearly there are some management advantages to designing the production system that maintained high calcium availability with consequent higher productivity. This would result in more of the nutrient outputs leaving as product and less as aqueous losses in stream water.

Gaseous Cycles

Gaseous global cycles do not have an important sedimentary phase. Thus rock weathering is not an important source of these elements. Again, each cycle has its own peculiarities, but nitrogen (N) is a good nutrient to illustrate a gaseous nutrient cycle. Nitrogen is also a limiting nutrient element in many cases and thus serves to illustrate the relationship between nutrient cycling and sustainability.

In a simplified model of the nitrogen cycle, the large reservoir of nitrogen is the atmosphere, in contrast to calcium, which has a huge reservoir in rocks. Like the calcium in rocks, nitrogen, as N_2 in the atmosphere, is unavailable to life in that form. The entry of nitrogen into life cycles requires capture and conversion of nitrogen into forms available to living organisms (nitrogen fixation). This fixation can happen through either abiotic processes (e.g., lightning) or biotic processes (biological fixation, e.g., through specially adapted symbiotic organisms like *Rhizobium* and *Frankia*). While common members of the community in some ecosystems, nitrogen fixers are lacking in many ecosystems. The other major source of nitrogen in terrestrial ecosystems is precipitation. Nitrate (NO_3^-) and ammonium (NH_4^+) are found in precipitation. Natural levels of these compounds are generally low, but industrial pollution has increased the level of these and related compounds to many times the natural levels. The net input rate of nitrogen to most terrestrial ecosystems is still rather low however. Once accumulated in terrestrial ecosystems through fixation or precipitation, nitrogen must be conserved through limiting losses, or the stock of nitrogen in the ecosystem will decline. Thus most natural ecosystems have evolved mechanisms to limit nitrogen losses. Again using a mass balance approach, the following simplification approximates the nitrogen balance in ecosystems:

precipitation + fixation + depletion of nutrient stock
= harvest + losses to ground- and surface water
+ atmospheric losses

In the absence of nitrogen fertilization, human management of nitrogen must then be designed to balance output of nitrogen through harvest removals and land management with inputs of nitrogen. With the primary input of nitrogen being bulk precipitation for many ecosystems, this rate of input generally translates to long rotation times (e.g., in the case of forestry). For annual agricultural crops where the annual removal of nitrogen in crops is high, this generally means that some kind of inorganic or organic fertilizer is required to replenish the stock of nitrogen in the soil. Depletion of nutrients through poor land management can also be an important management issue as discussed earlier. If erosion and leaching losses of nitrogen are high, sustainable practices require that harvest outputs be low in order to allow for a balanced nutrient cycle.

Contaminant Biogeochemical Cycles

The biogeochemistry of contaminants in ecosystems is of great concern to both environmental scientists and regulatory agencies. Many long-term, persistent contaminants have chronic effects on human health and also may lead to mortality through cancers and other deadly diseases. It is difficult to generalize about contaminant biogeochemistry because the elemental dynamics are so variable (e.g., mercury versus lead versus synthetic organics like DDT). For example, mercury (Hg) is volatilized (turns to gas and is then released into the atmosphere) in coal combustion, enters terrestrial and aquatic ecosystems through precipitation, and then re-enters the atmosphere through volatilization of gaseous mercury. While the accurate quantification of mercury "cycling" (as discussed earlier, most of many elements may actually not complete a cycle) is still under development (see Grigal 2002), one of the concerns is that human alteration of natural mercury cycling is resulting in both higher flows of mercury through terrestrial ecosystems and greater accumulation of mercury in aquatic ecosystems. Also, because mercury bioaccumulates (i.e., is concentrated to very high levels as it moves through the food chain) in aquatic ecosystems, it has critical impacts on aquatic organisms and the humans who eat fish from the sea and many lakes.

Tracing contaminant movement through all parts of the ecosystem and quantifying how and where it accumulates is a key part to understanding contaminant biogeochemistry and consequently managing contaminant dynamics to the greatest extent that we can. Unfortunately, because of the large number of industrial contaminants created by human activity, this work is very challenging and will require increased and continuing funding to achieve a level of understanding that can accurately inform sustainable management and policy development.

Managing Biogeochemical Cycles

While the framework of balancing inputs and outputs to maintain available pools of nutrients is simple in concept and helpful in evaluating sustainability, uncertainty in measuring all components of the cycle results in some large ranges in possible outcomes, especially when extrapolated over longer time frames. Long-term ecological research is just beginning to approach the time spans under consideration in many management scenarios (e.g., forest rotations). Nevertheless, current levels of accuracy in measuring important ecosystem variables—such as nutrients in bulk precipitation, weathering rates, denitrification, nitrogen fixation, outputs of nutrients in stream water, and stocks of macronutrients in soils—can provide important insights into key ecosystem cycles that can inform sustainable management. In particular, advances in research on the cycling of nitrogen, phosphorus, potassium, calcium, magnesium, and sulfur, and contaminants like lead and mercury can aid in framing decisions about sustainable practices and policies.

Deane WANG
University of Vermont

See also in the *Berkshire Encyclopedia of Sustainability* Eutrophication; Food Webs; Groundwater Management; Microbial Ecosystem Processes; Mutualism; Natural Capital; Nitrogen Saturation; Pollution, Nonpoint Source; Pollution, Point Source; Safe Minimum Standard (SMS); Soil Conservation

FURTHER READING

Bormann, Bernard T., et al. (1998). Rapid, plant-induced weathering in an aggrading experimental ecosystem. *Biogeochemistry, 43*(2), 129–155.

Bormann, F. Herbert, & Likens, Gene E. (1967). Nutrient cycling. *Science, 155*(3761), 424–428.

Bormann, F. Herbert; Likens, Gene E.; Fisher, D. W.; & Pierce, Robert S. (1968). Nutrient loss accelerated by clear-cutting of a forest ecosystem. *Science, 159*(3817), 882–884.

Grigal, D. F. (2002). Inputs and outputs of mercury from terrestrial watersheds: A review. *Environmental Review, 10*(1), 1–39.

Hemond, Harold F. (1980). Biogeochemistry of Thoreau's Bog, Concord, Massachusetts. *Ecological Monographs, 50*(4), 507–526.

Likens, Gene E. (2010). *Biogeochemistry of inland waters.* New York: Academic Press.

Likens, Gene E., & Bormann, F. Herbert. (1995). *Biogeochemistry of a forested ecosystem.* New York: Springer-Verlag.

Myers, Norman. (1980). *Conversion of tropical moist forests: A report prepared by Norman Myers for the Committee on Research Priorities in Tropical Biology of the National Research Council.* Washington, DC: National Academy of Sciences (NAS).

Stark, Nellie. (1971a). Nutrient cycling. I. Nutrient distribution in some Amazonian soils. *Tropical Ecology, 12,* 24–50.

Stark, Nellie. (1971b). Nutrient cycling. II. Nutrient distribution in Amazonian vegetation. *Tropical Ecology, 12,* 177–201.

Stark, Nellie. (1978). Man, tropical forests, and the biological life of a soil. *BioTropica, 10*(1), 1–10.

Vitousek, Peter M., & Matson, Pamela A. (2009). Nutrient cycling and biogeochemistry. In Simon A. Levin (Ed.), *Princeton guide to ecology* (pp. 330–339). Princeton, NJ: Princeton University Press.

Ocean Resource Management

The ocean, provider of food, oxygen, even medicine for millennia, is no longer the unchangeable body it was once thought to be. It is heavily affected by human activities, from overfishing to waste runoff to mineral extraction, destroying habitats and irrevocably changing ecosystems. National governments and international organizations must strengthen and integrate their efforts toward sustainability to retain the viability of marine ecosystems and resources.

The ocean—covering 72 percent of Earth's surface—provides vital, life-sustaining services to the global population. The world's oceans generate half the oxygen on Earth, are the primary regulator of global climate, and provide economic and environmental services to billions of people. The oceans also act as an important sink, having absorbed over 80 percent of the excess heat and approximately one-third of all anthropogenic carbon dioxide (CO_2) since the onset of the Industrial Revolution. Marine biodiversity and ecosystem resources and services provide basic life necessities, including food, fresh water, wood, fiber, genetic resources, medicines, and cultural products. Coastal areas are heavily utilized—half the world's population lives within 100 kilometers of the sea, and three-quarters of all large cities are located on the coast (UNEP and UN-HABITAT 2005).

Throughout history, people have had a long-standing connection with, and dependence on, the sea. Humanity's relationship with the ocean has changed in recent decades, increasing the demands and impacts placed upon the ocean. Today, the greater ocean no longer seems so vast and unreachable. It has also become apparent that the marine environment is not unaffected by our actions, even distant actions, on land and at sea, but that our compounding impacts have far-reaching consequences for often fragile marine ecosystems. The ocean is not the resilient body we once thought it to be but is heavily impacted by human actions throughout all regions of the world, as well as at great depths.

Human Activities Affecting the Ocean

Humans interact with the ocean in numerous ways. The most intense uses of ocean resources and space are often associated with more traditional activities, such as fishing, shipping, offshore oil and gas extraction, laying of cables, tourism and recreation, and coastal development. More recently, human activities have expanded to include mineral extraction, the extraction of marine genetic resources, marine renewable energy, the construction of artificial reefs, land reclamation, coastal defense, and dredging and dumping. The United Nations Conference on the Human Environment recognized in 1972 the overexploitation of living marine resources, the physical alteration of habitats, and marine pollution. These major threats to the ocean persist today.

Global Distribution and Impacts

Fishing is perhaps the most long-standing and common use of ocean resources. It is also the activity that most heavily exploits ocean resources. The United Nations Food and Agriculture Organization (FAO) estimates that fish provide over 3 billion people with 15 percent of their total animal protein intake, and some type of fishing activity occurs in every region of the ocean. Global decline in fish stocks began in the early 1990s. Today, with total catch exceeding 100 million tons per year (including discards, bycatch, and illegal, unregulated, and unreported fishing), the FAO estimates that 85 percent of stocks are either fully exploited, overexploited, depleted, or recovering from depletion, giving cause for concern about the long-term sustainability of these key resources (FAO 2010).

Destructive fishing practices also damage marine habitats. For example, bottom trawling, which involves dragging large, heavy nets along the sea floor, effectively decimates benthic habitats, that is, habitats of organisms living at the bottom of the ocean. In the deep ocean, huge reservoirs of marine biodiversity exist, particularly surrounding seamounts, hydrothermal vents, methane seeps, and deepwater corals. Organisms that thrive in these environments evolved under extreme environmental parameters and offer genetic material highly valuable for scientific discovery and commercialization. This material has numerous applications in pharmaceutical, biotechnological, and cosmetic fields. Many of these ecosystems are vulnerable to destructive fishing practices and potential overexploitation from commercial extraction.

The physical alteration and destruction of habitats is arguably the most important threat to coastal resources and environments. Social and economic development in these areas has led to coastal habitat destruction resulting from increasing pressures from population, urbanization, industrialization, marine transportation, and tourism. Such destruction often comes at significant environmental and economic (in terms of losses of ecosystem services) costs. For example, coral reefs, which provide habitat for over 1 million aquatic species, including thousands of fish species, and provide natural barrier protection from increased storm surges and wave activity, are estimated by the Economics of Ecosystems and Biodiversity project to be valued between US$130,000 and US$1.2 million per hectare per year (Diversitas 2009). Yet an estimated 58 percent of coral reefs worldwide are threatened, with habitat destruction one of the key contributors, though climate change and ocean acidification threaten far more. Mangroves are also highly valuable coastal ecosystems, providing protection against storms, nursery groups for offshore fisheries, and wood and non-wood forest products, with a total monetary value estimated at US$10,000 per hectare per year (Costanza et al. 1997), not accounting for additional services provided by these habitats (e.g., carbon sequestration). In the past century, over one-half of all mangrove forests have been lost, largely as a result of physical alteration. Wetland and sea-grass communities are also at risk and continue to decline worldwide, drastically reducing their ability to provide similar ecosystem services.

Marine pollution also results from human interactions with ocean resources and space. Approximately 90 percent of world trade is carried by ship. Shipping damages the marine environment through, among other activities, oil spills and accidental discharges, chemical accidents at sea, waste disposal and water pollution, sound pollution, and the discharge of ballast water—ships' ballast water transports approximately three thousand species of plants and animals each day, which can lead to an uncontrollable growth of invasive species in some marine ecosystems.

Notably, however, the most significant threat of marine pollution comes not from activities at sea but on land. Land-based activities contribute some 80 percent of all pollution entering the oceans. Sewage continues to be the largest source of contamination by volume, but wastewater and agricultural nutrient runoff are also large polluters. Together, excessive nutrients from sewage outfalls and agricultural runoff have contributed to a rise in the number of dead zones (hypoxic or anoxic areas) in the marine environment, from 149 in 2003 to over 200 in 2006, resulting in the collapse of some ecosystems (Nelleman, Hain, and Alder 2008). Plastics and other debris that make their way to the ocean accumulate and further affect marine resources and ecosystems. While it is difficult to calculate the distribution of waste in surface waters, water columns, and on the seafloor, recent studies and observations confirm that debris is transported by ocean currents and tends to accumulate in a limited number of convergence zones, or gyres, in what has been termed "garbage patches." For example, in the North Atlantic and Caribbean convergence zone, over 200,000 pieces of plastics per square kilometer have been found (UNEP 2011). Additional patches have been confirmed in an area midway between Hawaii and California, around the North Pacific Subtropical High, as well as off the coast of Japan in a small recirculation gyre. Another area of high concern is the North Pacific Subtropical Convergence Zone, where a high degree of marine debris concentrates (and where a high diversity of marine life also exists). One model simulation of global marine litter distribution after ten years shows plastics converging in the five gyres, namely the Indian Ocean, North and South Pacific, and North and South Atlantic (IPRC 2008). The debris has lethal and sublethal effects on biodiversity, entanglement, chemical contamination, and the alteration of community structures. In a recent study of planktivorous fish from the North Pacific gyre, an average 2.1 plastic items were found per fish (Boerger et al. 2010). The global community needs to study, better understand, and address the potential impacts of accumulation and releases from plastic particles, including persistent, bioaccumulating, and toxic substances. Enhanced understanding is needed as well on the long-term impacts and effective strategies for addressing other types of hazardous substances leeching into the marine environment, such as mercury, lead, polycyclic aromatic hydrocarbons (PAHs), and polychlorinated biphenyls

(PCBs), which also are persistent, toxic, and bioaccumulate in fish, shellfish, and other marine organisms.

Management Challenges

The overarching framework for global ocean management is the 1982 United Nations Convention on the Law of the Sea (known as UNCLOS). Specific management provisions and challenges vary across sectors and geographic areas and through time. Managing human interactions with respect to ocean resources means balancing environmental and developmental needs. In parts of the world where persistent poverty and inequality loom large, strategies must focus on the longer-term benefits of sustainable management practices.

Inherent inequities exist in the global trade of ocean products, and there is no sound framework for benefit sharing. Sixty-four percent of the global ocean is beyond national jurisdictions, and there is no clear international framework governing the exploitation of high-seas marine resources and the protection of the marine environment. The global financial recession has complicated the picture by rearranging national priorities and capacity; sustainable practices compete with priorities in agriculture, infrastructure, energy, health, and education.

Global climate change increases the vulnerability of ecosystems and coastal populations, especially the world's poor. For the millions of people and local economies highly dependent on marine resources, climate variations increase poverty and food insecurity and lead to loss of livelihood and living space.

International, regional, and national governance issues pose major barriers to the sustainable ocean and coastal agenda. In many countries ocean resource management agencies are chronically underfunded and understaffed. Even in countries with strengthened institutions, addressing management challenges under national jurisdiction in the 200 nautical-mile exclusive economic zone requires expertise, equipment, and vessels for monitoring, control, and surveillance. A high level of technology, capacity, and coordination, supported by broader legal and institutional frameworks at the international and national levels, are needed to address multiple uses and expectations in ever-more-crowded oceans and coasts.

Management of Sustainable Ocean Governance

To overcome obstacles toward sustainable ocean governance and address increasing resource and user conflicts in ocean areas, national governments and international authorities recognize the need to adopt approaches for integrated coastal and ocean management and ecosystem-based management, approaches that shift focus from managing specific, single-sector marine uses to managing multiple uses on an ecosystem basis. The 2002 World Summit on Sustainable Development (WSSD) called for the application of the ecosystem approach by 2010 and the promotion of integrated coastal and ocean management at the national level. These paradigms realize the interrelations of marine ecosystem services and seek to implement holistic, sustainable management and governance.

The WSSD also agreed to achieve a significant reduction by 2010 in the current rate of biodiversity loss at the global, regional, and national levels as a contribution to poverty alleviation, as well as the establishment of marine protected areas, consistent with international law and based on scientific information and including representative networks, by 2012. It further called for the development of diverse approaches and tools, with a focus on the ecosystem approach and the elimination of destructive fishing practices. The 2006 Eighth Conference of the Parties to the Convention on Biological Diversity provided further clarity to marine biodiversity targets, calling for the effective conservation of at least 10 percent of each of the world's marine and coastal ecological regions and for the protection of particularly vulnerable marine habitats, such as tropical and cold-water coral reefs, seamounts, hydrothermal vents, mangroves, sea grasses, spawning grounds, and other vulnerable marine areas.

Today, however, only about 1 percent of the world's oceans have been afforded any protection, and renewable marine resources continue to be depleted. Strengthened efforts toward improving international coordination and national capacity are needed to move toward a more integrated approach. Marine protected areas are a useful tool in this approach, though they are but one of the measures needed for sustainable governance. For example, marine spatial planning optimizes the use of marine space to benefit economic development and the marine environment by balancing sectoral interests and the sustainable use of ocean resources, and is emerging as an important decision-making tool. Ecosystem valuation is also critical, providing analyses of the full economic valuation of, for example, coral reef ecosystems. Such valuation estimates both market and nonmarket goods and services to promote better understanding of the economic and societal consequences of environmental change.

Governance gaps and institutional deficiencies related to ocean resource management must be addressed in the coming years. International cooperation, compliance, and enforcement mechanisms must be enhanced. The United Nations Conference on Sustainable Development, held in Rio de Janeiro, Brazil, in June 2012 (Rio+20, being held twenty years after the first Rio Conference) represents a unique opportunity for the international community to secure renewed political commitment for the sustainable management of ocean resources, assess progress to date, and review remaining gaps in implementation.

Though no common approach effectively addresses all ocean resource management issues, decision makers must pursue strategies fully supported by common tools and techniques, independent science, monitoring and assessment, sustainable finance mechanisms, and methods of evaluation. The three pillars of sustainable development—economic development, social development, and environmental protection—cannot be achieved without the sustainable management of ocean resources. Coordinated and proactive measures must be pursued to ensure the viability of marine ecosystems and resources, thus securing the continued life-support functions humanity receives from the global oceans.

Kateryna M. WOWK
Global Ocean Forum

See also in the *Berkshire Encyclopedia of Sustainability* Best Management Practices (BMP); Catchment Management; Coastal Management; Fisheries Management; Large Marine Ecosystem (LME) Management and Assessment; Marine Protected Areas (MPAs); Ocean Acidification—Management; Pollution, Nonpoint Source; Pollution, Point Source

FURTHER READING

Beck, Michael, et al. (2009). Shellfish reefs at risk: A global analysis of problems and solutions. Arlington, VA: The Nature Conservancy.

Boerger, Christiana; Lattin, Gwendolyn; Moore, Shelly; & Moore, Charles. (2010). Plastic ingestion by planktivorous fishes in the North Pacific Central Gyre. *Marine Policy Bulletin, 60*(12), 2275–2278.

Corcoran, Emily, et al. (Eds.). (2010). *Sick water? The central role of wastewater management in sustainable development: A rapid response assessment.* Arendal, Norway: United Nations Environment Programme (UNEP), UN-HABITAT, GRID-Arendal.

Costanza, Robert, et al. (1997). The value of the world's ecosystem services and natural capital. *Nature, 387*(6630), 253–260.

Diversitas. (2009, October 28). What are coral reef services worth? $130,000 to $1.2 million per hectare, per year. Retrieved August 2, 2011, from http://www.sciencedaily.com/releases/2009/10/091016093913.htm#

Food and Agriculture Organization of the United Nations (FAO). (2010). *The state of world fisheries and aquaculture report 2010.* Retrieved August 5, 2011, from http://www.fao.org/docrep/013/i1820e/i1820e.pdf

GESAMP (IMO/FAO/UNESCO-IOC/WMO/WHO/IAEA/UN/UNEP Joint Group of Experts on the Scientific Aspects of Marine Environmental Protection) & Advisory Committee on Protection of the Sea. (2001). *A sea of troubles* (GESAMP Reports and Studies No. 70). Arendal, Norway: United Nations Environment Programme, GRID-Arendal.

International Pacific Research Center (IPRC). (2008). Tracking ocean debris. *Climate, 8*(2), 14–16.

Nellemann, Christian; Hain, Stefan; & Alder, Jackie. (Eds.). (2008). *In dead water: Merging of climate change with pollution, over-harvest, and infestations in the world's fishing grounds.* Retrieved July 28, 2011, from http://www.unep.org/pdf/InDeadWater_LR.pdf

United Nations Conference on the Human Environment. (1972). Declaration of the United Nations Conference on the Human Environment. Stockholm, 5–16 June 1972. Nairobi, Kenya: United Nations Environment Programme (UNEP).

United Nations Development Programme (UNDP). (2002). *Conserving biodiversity, sustaining livelihoods: Experiences from GEF-UNDP biological diversity projects.* Retrieved August 2, 2011, from http://www.undp.org/gef/new/BiodiversityBrochure.pdf

United Nations Environment Programme (UNEP). (2011). *UNEP Year Book 2011: Emerging Issues in our Global Environment.* Retrieved August 2, 2011, from http://www.unep.org/yearbook/2011/

United Nations Environment Programme (UNEP). (2007). *Global environment outlook 4: Environment for development.* Valletta, Malta: United Nations Environment Programme. United Nations Environment Programme (UNEP) & United Nations Human Settlements Programme (UN-HABITAT). (2005). *Coastal area pollution: The role of cities.* Retrieved August 2, 2011, from http://www.unep.org/urban_environment/PDFs/Coastal_Pollution_Role_of_Cities.pdf

Permaculture

A permanent agriculture underpins a permanent culture. While permaculture is often thought of as a gardening system, it also encompasses the design of urban areas and ecovillages. Advocates of permaculture argue that most food, water, and shelter requirements can be met from local sources, from systems powered mainly by renewable energy. With limited resources globally, permaculture is a significant tool for the design of sustainable human habitats.

Permaculture is an integration of many skills and disciplines, brought together to design ways of living sustainably in the twenty-first century. The essence of permaculture is ancient in origin—taking inspiration from those civilizations of the world that have survived for thousands of years. The modern permaculture movement emerged as a response to the oil crises of the mid-1970s and continued to respond to various later environmental crises. Permaculture has broadened from its origins in designing sustainable landscapes to be recognized as a tool to help create sustainable societies. It has been used to design ecovillages, community gardens, and city farms. A global network of permaculture activists demonstrates practical solutions through design of sustainable systems, developing a human-centered approach, in which community, economics, legal structures, and the built environment are included.

What Is Permaculture?

Permaculture's advocates start with the axiom that the Earth's resources are finite but that infinite external energy is available from the sun. From this follows that the best way to convert sunlight into something people can use is by growing plants. A given area of perennial plants will return food, fiber, and timber. The same area of solar panels could generate more energy, but unlike solar panels, the plants are low cost and can be grown by most people.

Perennial plants are especially useful because they don't need as much labor as annual crops do and because they use less fossil fuel to grow than monoculture crops such as wheat, rice, maize, and potatoes. The emphasis on perennial plants provided the "perma"(nent) aspect of the name *permaculture*.

Because humans need more than food and water to survive, permaculturists have developed seven principles, or domains, of sustainable design to guide development of larger systems:

- land and nature stewardship
- built environment
- tools and technology
- culture and education
- health and spiritual well-being
- finance and economics
- land tenure and community governance

Permaculture design gives priority to trees and forests. These long-lived, self-sustaining systems are essential to life on Earth. Through photosynthesis, trees naturally turn the energy of sunlight into food and energy for themselves and wood, fruit, medicine, and fiber that people can harvest, and at the same time they stabilize the landscape. They have had millions of years to perfect this process.

Permaculture design promotes the replication of natural systems to create human-made systems. These systems have to encompass social and economic systems that sustain humans, as well as systems that produce food, fiber, timber, and clean water and air.

Permaculture gardens or field crop areas are designed to make best use of relationships between species of plants and animals and to have multiple types and levels of production. Although some plants don't grow well in close proximity, most can be planted near each other, and some have beneficial interactions. For example nitrogen-fixing bushes are planted around fruit trees—close enough that their roots intertwine. When the nitrogen-fixing bush is pruned or browsed, some of the nitrogen around the roots is given up and made available to the fruit tree, reducing or avoiding the need to apply fertilizer (Holmgren 1996, 46)

By including plants that grow at different heights, harvest can be obtained from various parts of the garden simultaneously. In a densely planted forest garden, root crops, ground cover, soft fruit from bushes, pome fruit from trees, and berries from vines can all be gathered from a small orchard planted with a variety of species. The soft berry fruit and ground cover such as strawberries are already adjusted to lower light conditions, so they are not restricted by the shade of fruit trees, as long as water and nutrients are available (Hart 1996).

In a permaculture design, both the soil and tanks or dams are used to make sure that there is adequate availability of rain water. Nutrients come from leaf litter and microbial activity—just as in any organic garden—and from making compost. With good fencing and management, animals can be beneficially introduced, and their manures can be composted or used directly by the plants when animals are introduced into the garden. Chickens under fruit trees is one example, but ducks, geese, sheep and pigs can all be used, from time to time, in permaculture systems to assist with nutrient cycling, control of grass, weeds or pests, and for their inherent products such as meat, milk, and eggs. The website of the Hunter & Central Coast Regional Environmental Management Strategy (HCCREMS 2010), a framework developed to address a range of environmental issues within New South Wales, Australia, hosts fact sheets full of information about the composting process.

This conscious design that links plants, animals, and structures such as water tanks has a natural appeal to the homesteader and hobby farmer but has been taken up on a broader scale, for example, where grain farmers and graziers increase the number of hedges and shelterbelts of trees, and where orchardists introduce animals to help manage their fruit crops (Mollison 1988, 60–61; Lillington 2007, 102–104).

The scientific basis of permaculture comes from both natural systems ecology (an ecological approach to agriculture) and from thermodynamics. The work of the US ecologists Howard and Elisabeth Odum, who incorporated the laws of thermodynamics and extended the concept of embodied energy, significantly influenced the Australian permaculture pioneer David Holmgren. In 2001, the Odums published *A Prosperous Way Down: Principles and Policies*, which proposes ways for the human race to better understand energy—where we get it from, how we use it, and what happens once we have used it—and then to be able to design sustainable systems. In particular, the Odums and permaculture practitioners point to the need to recognize and act upon the fact that people currently rely almost entirely on fossil fuel energy, which is both polluting and finite.

History and Development

The early thinking about permaculture combined ecology, landscape, and agriculture. In the 1970s, the Australian academics David Holmgren and Bill Mollison collaborated on what was to become permaculture. David Holmgren's graduate thesis became the basis for the book *Permaculture One*, published in 1978. This book showed how ecology and agriculture could be combined, by conscious design, to create a landscape filled with sustainable food production systems.

David Holmgren describes that time:

Permaculture arose from interaction between myself and Bill Mollison in the mid 1970s. We were two (very different) social radicals on the fringes of (different) education institutions, at the global fringes of western industrial society in Tasmania. Bill Mollison as bushman turned senior tutor, in the Psychology Dept.

of the Tasmanian University, attracted large student audiences to hear his radical and original (prepermaculture) ideas while outraging the academic establishment.

I was a student in the Environmental Design School, a revolutionary "experiment" in tertiary education at the Tasmanian College of Advanced Education. This design school ran for ten years under the inspired leadership of Barry McNeil, a Hobart architect and education theorist. There was no fixed curriculum but a strong emphasis on decision making processes and problem solving. Self assessment, democratic organization and many other elements which radicals within tertiary institutions only dream about were reality within the school. (Holmgren 2006)

An Ethical Approach

As the world's population diversifies from a fossil fuel economy to a renewables-based economy, many people are seeking rules or guidelines. Permaculture has always been an explicitly ethical approach. These ethics are not unique to permaculture and are similar to the precautionary principle (the idea that if the consequences of something— such as the use of nanotechnology— are unknown, it is best to study the issue before acting). As they studied early permaculture societies, Bill Mollison and David Holmgren observed similar ethical bases in them. These ethics are usually expressed as care for the Earth, care for people, and fair shares. (*Fair shares* captures the idea that no one is hoarding and that there are limits to consumption and population growth.)

Care for the Earth implies that even as they use more of the sun's energy, humans need to be careful with all the other resources of the Earth. People need more than energy—they also need forests, fish, good soil, minerals, and many other raw materials. The second and third ethics can be seen as derived from the first. Care for people and sharing surpluses spring from the understanding that abundant solar energy is useless unless there is a permanent civilization and a healthy, living planet for plants and animals, including humans.

Robert Hart, another ecological leader of the twentieth century whose work paralleled the development of modern permaculture, also emphasized ethical principles. He developed numerous ethical and sustainable projects and the idea that landscapes can be designed as edible food forests. In his book *Beyond the Forest Garden*, Hart (1996) wrote, "With our present knowledge, there is no technical reason why every woman, man and child on Gaia's earth should not be adequately fed, clothed, housed and given the opportunity for self realization."

Principles

The book *Introduction to Permaculture*, by Bill Mollison, built on earlier work to codify the basic principles of permaculture. David Holmgren's subsequent *Permaculture: Principles & Pathways beyond Sustainability* lists twelve permaculture principles and adds the seven domains of sustainable design. The seven domains clarify that permaculture is more than organic gardening. Permaculture design principles are intended to assist in creative thinking— developing integrated solutions where landscape, buildings, people, commerce, technology, health, education, and governance come together.

Permaculture's ethics and principles are considered when designing for any situation. In contrast, permaculture techniques and strategies have to be applied thoughtfully to each site-specific situation.

Applied Permaculture

People who study permaculture generally go on to take action, but the actions are of many kinds. Some simply grow more food, others design and develop privately owned land with energy-efficient housing and self-sustaining landscapes. Some develop ecovillages or

teach courses in permaculture. Others have developed the *permablitz*—an intensive working party of local volunteers to turn an unused or weedy garden into a food-producing system—and still others have gone on to establish or support city farms and community gardens.

Permaculture teachers and activists encourage others to examine where their food and essential needs come from and to measure (audit) their use of resources. Students of permaculture are encouraged to set targets to reduce their fossil fuel use. This is often done through growing at least some of the food they consume and obtaining most of their food from local and in-season sources to reduce the distance their food travels to their tables.

An audit is a simple self-check—for example, reading the household's electric, water, and gas meters and setting targets to reduce the amount used or noting the distance driven by car each week and seeking to reduce it. Through these observations, individuals and households better understand and reduce their energy use. Energy is most obviously measured at the electric or gas meter or at the gas pump, but "waste" leaving a home, office, or factory is also a form of energy, and this can be measured as well. Waste is an unused resource—whether it is the very visible weekly household garbage collection or the unseen by-products of manufactured goods. Permaculture design seeks to turn any waste product (an output) into a resource (input) that can be used elsewhere in the system. Backyard chickens are symbolic of permaculture, because they provide important services to people as they convert kitchen scraps and garden waste into eggs, meat, and feathers. Because reducing the distance driven in a private car is a way to improve the local and global environment, permaculture designers aim to create livable cities and towns where more trips can be made by foot, bike, or public transportation, and where fewer trips need to be made in total as basic needs can be met closer to home.

Garden Agriculture

Although permaculture is not simply a gardening system, gardens are often the way that people become more conscious of the need for permaculture's holistic approach to design. A significant part of being able to sustain our population is for each suburban garden to produce food or fiber, so that whole cities become a kind of farm. Many private gardens in every part of the world contain food plants that are maintained for enjoyment as well as production. David Holmgren and others call this *garden agriculture*. Redirecting the time and resources devoted to ornamental gardens into growing edible plants can achieve high levels of local production. Human labor

rather than machines provide the major power input. There is room in every garden for some annual vegetables, and a large proportion of a household's fresh food needs can be grown in a small area as long as the soil is fertile.

Transitioning to Resilience

From about 2002 onward, permaculture practitioners were drawing attention to the "peaking" of many key resources, such as oil and gas. American geo-scientist, M. King Hubbert (1903–1989) created and used the models behind peak oil in 1956 to accurately predict that US oil production would peak between 1965 and 1970 (American Heritage Center 2009). This model and its derivatives have described the peak and decline of oil production globally, and they have proved useful in forecasting the peak of other key resources, such as phosphorous (Heinburg 2007).

The first decades of the twenty-first century will be a period of transition to a wider range of energy sources. Permaculture principles are already assisting in the design for ways of living where energy comes from the renewables driven by the sun rather than from fossil fuels.

Permaculture was the starting point for the Transition Town movement that originated around 2005 in Ireland, for example. Those involved in the transition/permaculture movement have recognized that significant change is underway and that embracing change will allow people to be better prepared for an uncertain future.

Permaculture Education

Since the early days of permaculture, courses and workshops have spread permaculture's ethics, principles, and techniques. Many students who took courses in the 1980s went on to become teachers, and growth has been exponential. The globally recognized educational standard is the Permaculture Design Course (PDC), which consists of at least seventy-two hours of study taught by experienced permaculture teachers. Some parts of the course cover the generic aspects of permaculture; some parts are specific to the region in which the course is taught.

The PDC is unusual for two reasons. First, it is resilient: thousands of such courses have been run in dozens of countries over three decades without the support of an institutional or financial backer. Second, the qualification is widely respected, with hundreds of thousands of graduates, perhaps more than a million. The longevity of the PDC is testimony to the dedication of permaculture teachers and networks of permaculture activists who ensure that those who are "doing permaculture" are also the quality control mechanism for permaculture.

Although the PDC is not recognized by many traditional institutions in most parts of the world, a few do have accredited courses in permaculture. Australia, where a diploma of permaculture is part of the Australian Qualifications Framework, is one country that does support this kind of education.

Can Permaculture Feed the World?

Permaculture is sometimes dismissed as being impractical and too labor intensive. Some critics have questioned whether there is enough scientific data to validate claims that a system of perennial plants will yield more per acre than a single crop such as wheat, maize, rice, or soy. In a 2001 article, Greg Williams, editor of *HortIdeas* magazine, presents a challenge to the argument that perennial is best. Williams argues that a garden based on a meadow is more productive than one based on a forest. Williams does not take account of extra energy that is needed in the annual immature system of the meadow compared with the perennial mature forest garden approach, however, or the soil loss that occurs whenever tillage happens.

Instead of debating what can be produced from a hypothetical acre, permaculturists prefer to assess how much food a town or city needs and whether it can be produced in or near that urban area. This is one of the reasons permaculture designers focus on cities as well as rural areas—there is much unproductive open space in the city that could be converted into food production.

Advocates of permaculture argue that to eat sustainably, you have to know how much energy goes into the food through the use of tractors, trucks, processing, and packaging. Most foods based on grains are very energy intensive. Permaculturists suggest that people do not need to rely on these annual crops; they simply have to work out what types of food they want to eat and ask if they can be produced in their own garden. Not all food can be produced from a forest garden—most permaculture practitioners who set out to grow a substantial amount of their own food end up with a mix of annual vegetables and tree crops. Beyond that, purchases in the permaculture system come from local growers in preference to food that has traveled a long distance. To be successful in permaculture, people are likely to change their diets substantially.

In addition, some researchers are concerned that permaculture advocates the spread of weeds. Permaculturists, however, stress the use of native plants whenever possible and express concern that genetically modified crops are more dangerous than any weed. Other practitioners say that modern agriculture has damaged the Earth to such an extent that including non-native plants to provide for a sustainable future is more important than preserving current ecosystems.

Toward a More Sustainable Society

Many contemporary thinkers and writers, including permaculturists, are convinced that there has to be some kind of descent from the heights of mass consumption. Since 1700, and especially since 1950, human population has grown enormously, along with a parallel growth in the consumption of the Earth's natural resources—fossil fuels like oil, coal, and gas, and natural resources like fish, forests, and topsoil. This economic and population growth is a typical response of any group of plants, animals, or bacteria that has access to plenty of food and energy (WRI 2009).

Growth will, however, reach a limit. On a planet with finite physical resources, people will have to make smarter use of renewable energy sources, most of which originate with the sun. The world is constantly growing and changing. The years since 1900 can be compared with climbing a mountain—lots of energy is needed and new tasks have to be achieved at every step. At the top of the mountain, the peak, the climber has to find the way back down, which can often be the more dangerous part.

The sun continues to offer a significant and abundant alternative source of energy, but it comes in a different form from the concentrated supplies of oil, gas, and coal. Permaculture practitioners have shown how people can live in a renewable system powered by sun (the solar economy) rather than oil (the fossil fuel economy).

Ian R. LILLINGTON
Swinburne University, Australia

See also in the *Berkshire Encyclopedia of Sustainability* Agricultural Intensification; Agroecology; Carrying Capacity; Home Ecology; Hydrology; Landscape Architecture; Resilience; Soil Conservation; Urban Agriculture; Urban Vegetation; Water Resource Management, Integrated (IWRM); Urban Agriculture; Urban Forestry

FURTHER READING

American Heritage Center. (2009). M. King Hubbert papers. Retrieved November 30, 2011, from http://digitalcollections.uwyo.edu:8180/luna/servlet/uwydbuwy~62~62

Dawborn, Kerry, & Smith, Caroline. (Eds.). (2011). *Permaculture pioneers: Stories from the new frontier.* Hepburn, Australia: Melliodora Publishing.

Giono, Jean. (1953). *The man who planted trees.* New York: Reader's Digest.

Hart, Robert. (1996). *Beyond the forest garden.* London: Gaia Books.

Heinburg, Richard. (2007). *Peak everything: Waking up to the century of declines.* Sebastopol, CA: Post Carbon Institute.

Heinberg, Richard, & Bomford, Michael. (2009). *The food and farming transition: Toward a post-carbon food system.* Sebastopol, CA: Post Carbon Institute.

Heinburg, Richard. (2011*). The end of growth.* Sebastopol, CA: Post Carbon Institute.

Holmgren, David. (1996) *Melliodora (Hepburn permaculture gardens): A case study in cool climate permaculture 1985–2005.* Hepburn, Australia: Holmgren Design Services.

Holmgren, David. (2002). *Permaculture: Principles and pathways beyond sustainability.* Hepburn, Australia: Holmgren Design Services.

Holmgren, David. (2006). *Collected writings & presentations 1978–2006.* Hepburn, Australia: Holmgren Design Services.

Holmgren, David. (2009). *Future scenarios.* Hepburn, Australia: Holmgren Design Services

Hopkins, Rob. (2011). *The transition companion: Making your community more resilient in uncertain times.* White River Junction, VT: Chelsea Green Publishing.

Hunter & Central Coast Regional Environmental Management Strategy (HCCREMS). (2010). Factsheets for environmental education. Retrieved November 30, 2011, from http://hccrems.interesting.com.au/HCCREMS-Resources/Resource-library/Education/HCCREMS-Fact-Sheet-Source-list-for-educators.aspx

Lillington, Ian. (2007). *The holistic life: Sustainability through permaculture.* Adelaide, Australia: Axiom Press.

Mollison, Bill. (1979). *Permaculture two.* Ealing, UK: Corgi Press.

Mollison, Bill. (1988). *Permaculture: A designer's manual.* Tyalgum, Australia: Tagari Publications.

Mollison, Bill, & Holmgren, David. (1978). *Permaculture one.* Ealing, UK: Corgi Press.

Odum, Howard, & Odum, Elisabeth C. (2001). *A prosperous way down: Principles and policies.* Boulder: University Press of Colorado.

Permaculture Australia. (n.d.). Accredited Permaculture Training (APT) course accreditation documents. Retrieved November 17, 2011, from http://permacultureaustralia.org.au/2011/11/17/apt-course-accreditation-documents/

Seuss Geisel, Theodor. (1971). *The Lorax.* New York: Random House.

Williams, Greg. (2001) Gaia's garden: A guide to home-scale permaculture. Retrieved November 30, 2011, from http://findarticles.com/p/articles/mi_m0GER/is_2001_Winter/ai_81790195/?tag=content;col1

World Resources Institute (WRI). (2009). Population and consumption. Retrieved November 30, 2011, from http://earthtrends.wri.org/updates/node/360

Plant-Animal Interactions

Broadly defined, any relationship occurring between organisms in the kingdoms Animalia and Plantae is classified as a plant-animal interaction. Plant-animal interactions are common features of virtually every environment, including all marine, freshwater, and terrestrial biomes. Many such interactions demonstrate evolutionary principles and the myriad ways that species interactions influence the functioning of the biosphere.

Biologists and naturalists have long been fascinated with plant-animal interactions (PAI), the relationships between organisms in the kingdoms Animalia and Plantae. The seeming simplicity of the formulation conceals an enormous number and diversity of ecological relationships and fundamental processes, ranging from the obscure to the ubiquitous. As a result, there is an extensive history of investigation into these often fascinating relationships. The preeminent biologist Charles Darwin wrote extensively about PAI in *On the Origin of Species by Natural Selection*, although many PAI were the focus of a considerable amount of descriptive ecology prior to Darwin's text. Today, PAI remain a centerpiece within many of ecology's central theories, including, among others, coevolution and consumer-resource theory.

General Categories of PAI

Plant-animal interactions range from the general to those that are highly specific and involve elaborate evolutionary adaptations. An example of a general PAI is a plant that provides shelter for an animal, such as a tree that provides critical habitat for a nesting bird. Some animals are flexible in their choice of plants; in contrast, some insects are highly specialized, living and laying eggs on only one plant species. In an attempt to categorize and describe the plethora of PAI, biologists further categorize PAI as being commensal, in which one partner benefits while the other is unaffected; antagonistic, in which the interaction is detrimental to at least one partner; or mutualistic, in which both the plant and animal partners both benefit. Interactions are classified by whether an individual partner has more, less, or the same number of offspring as a result of the relationship, in terms of higher or lower fitness. Although the ultimate value is the reproductive success (fitness) of the interacting plants and animals, this can be quite difficult to measure. Thus other metrics such as photosynthetic carbon gain, growth rate, longevity, and survival are often used as surrogate estimates of fitness.

Commensal Interactions

Commensal plant-animal interactions, while straightforward in theory, are somewhat difficult to demonstrate. This is because there is always some question whether an interaction has a completely neutral effect on one of the species involved. Take the previous example of a bird nesting in a tree, which clearly benefits the bird but may or may not influence the tree. If the presence of the nest has no effect on the tree's growth and reproduction, then the relationship is truly commensal. The bird may eat herbivorous insects that feed on the tree, thus having a positive effect. Alternatively, the nest may block sunlight or weigh down branches away from sunlight exposure, thus having a negative effect. The task of conclusively demonstrating commensalisms in this type of interaction involves experimentally removing nesting birds from some trees and leaving others unchanged and comparing the fitness of the two groups.

Antagonistic Interactions

The most common plant-animal interactions are antagonistic and involve the direct consumption of plants by animals (called herbivores) for food. This general PAI serves as the fundamental process for transferring the energy of sunlight to the animal biomass in all ecosystems. Herbivores can be highly specialized or unselective generalists and span a huge range of body sizes, from tiny leaf-eating and sap-sucking insects to large herbivores such as elephants, or the selective Chinese giant panda whose diet consists almost entirely of bamboo. Herbivores have evolved a variety of feeding styles to consume plants. For example, insects in the order Hemiptera, such as aphids, leafhoppers, and scale insects, have piercing and sucking mouth parts specialized to suck fluids directly from the vascular system (xylem water and phloem sugars) of the plant. Other insects, such as those belonging to the orders Orthoptera (grasshoppers and crickets) and the larvae of Lepidoptera (moths and butterflies), have chewing mouthparts that allow them to bite and tear leaf material. Vertebrate herbivores also come in a variety of types and sizes, including fresh- and saltwater fish that feed on algae, small rodents that eat parts of leaves, and large-bodied mammalian herbivores that forage on woody plant species (called browsers) or that eat more ground-dwelling herbaceous plants (grazers). Large-bodied mammal herbivores have high-crowned teeth and specialized digestion to facilitate the internal decomposition of plant material. Some ecosystems throughout the world, such as Serengeti National Park in Tanzania, Africa, and Yellowstone National Park in Wyoming, are famous for the abundance and diversity of these large mammal herbivores (also termed megaherbivores). These ecosystems have been labeled grazing ecosystems or browsing ecosystems because of the large proportion of energy transferred from primary producers to the primary consumers, grazers, and browsers.

Plants have evolved a broad spectrum of defenses against herbivory, ranging from tolerance to resistance of defoliation. Herbivory-tolerant plants have high growth rates and are able to reallocate stored carbohydrates to defoliated stems rapidly. Additionally, plants tolerant of herbivory often have architectures that protect carbohydrate-rich storage organs, found below ground or out of the bite range of herbivores. Plants that are resistant to herbivory employ either structural or chemical defenses that deter or even harm herbivores. The most basic structural defense of plants is the production of cell walls and fibrous tissues composed of cellulose and lignin, a main component of wood, which are difficult for herbivores to chew and digest. More specialized structures include thorns, barbed spines, hooks, and hairs that protect especially the photosynthetic tissue of plants. Plant chemical defenses, also known as secondary compounds (or metabolites), are metabolic products not necessary for primary growth and reproduction. The chemistry of plant secondary compounds is complex but well studied because of the deep historical connection with humans. For example, plant secondary compounds are responsible for a rich array of chemicals used by humans, including herbal stimulants (coffee, nicotine), narcotics (cocaine), spices (nutmeg, mint), and a vast array of medicines that treat everything from headaches (aspirin from willow bark) to cancer (taxol from the Pacific yew tree).

Many small rodents and birds, known as granivores, consume seeds rather than plant tissues. Another group of specialized herbivores called frugivores feed specifically on plant fruit; this is a diverse group including a wide variety of insects, birds, and mammals. Although less well studied than their above-ground counterparts, there is also a diverse community of below-ground herbivores, composed of nematodes, insects, and rodents, that forage on plant roots.

Not all antagonistic relationships involve animals eating plants. One of the more interesting deviations from the typical pattern is that of the carnivorous plants. Currently there are more than six hundred species of carnivorous plants described, including the well-known Venus fly trap and pitcher plant, which trap and slowly extract nutrients from decomposing arthropods. The first popular scholarly text on carnivorous plants was written by none other than Charles Darwin in 1875.

Mutualisms

Plants and animals also engage in a wide diversity of interactions that benefit both partners. One ubiquitous example is pollination, in which animals feed on nectar and pollen from flowers, transferring pollen to other plants, the foundation of the highly successful sexual reproduction of flowering plants. The vast majority of pollinators are insects, but the group also includes birds, bats, rodents, monkeys, and even lizards.

A second type of common mutualism between plants and animals is seed dispersal. Animals benefit by consuming fruits that house the seeds, while plants benefit by having their seeds dispersed long distances by animals, thus increasing their offspring survival probability. Especially for large-seeded plants, long-distance dispersal of seeds would be physically impossible without animal vectors. In this regard, the human domestication of fruits and vegetables may represent one of the most extensive plant-animal mutualisms on Earth. Another type of mutualism involves animals that protect plants from other animal herbivores.

Ants and Acacias

The PAI involving ants and acacias is one of the best-known examples of mutualism. In tropical woodlands and savannas throughout the world, trees belonging to the genus *Acacia* produce hollow, swollen structures on their twigs that provide shelter for stinging ants. Moreover, these trees also have glands at the base of their leaves that secrete carbohydrate-rich nectar on which the ants feed. Thus, the ants benefit by receiving both a place to live and a source of energy-rich food. This relationship is mutualistic because the trees benefit in return: the ants swarm to attack leaf-eating mammal and insect herbivores. This relationship is very effective and even protects acacias from African elephants, the largest terrestrial herbivore on Earth. Interestingly, the ants do not attack bees that pollinate the acacia flowers because of a chemical released by the plant that somehow prevents ants from approaching during pollination.

Yucca and Yucca Moths

Another classic example of mutualism is the close association between yucca plants and yucca moths. This interaction is a highly specialized relationship in that each species of yucca has only a few, and sometimes just one, species of moth with which it interacts. Also, the yucca moths are entirely dependent on the yucca plants for their own reproduction, while yucca plants require cross pollination between different individuals and rely completely on the yucca moths for pollen transfer. The moths pack the pollen into the stigma of the yucca plant, ensuring fertilization. While adult moths do not feed, female yucca moths lay eggs in the flowers, and the emerging larvae then feed on the developing seeds. This relationship is classic in that it provides an example of extreme specificity between partners and a relationship in which mutualism and antagonism are balanced in a strong coevolutionary relationship.

Bees and Bee Orchids

A final example of PAI mutualism is the "deceptive" plant pollination that occurs between bees and bee orchids. Bee orchids have evolved a mechanism to deceive bees into pollinating their flowers through both visual and chemical mimicry of the female bee. The bee orchid produces floral structures that look like a female bee and emits volatile chemical compounds that mimic female reproductive pheromones. Bees are attracted to the plant and are deceived into "mating" with the flowers. In reality, the male bees distribute pollen between individual plants but receive nothing in return.

Coevolution and the Antiquity of PAI

The two most species-rich groups of macroscopic terrestrial organisms are plants and insects. One body of theory suggests that the global diversity of these groups is a consequence of their long coevolutionary history, which has persisted since the first invasion on land over 450 millions years ago. Evidence of ancient herbivory, in the form of fossilized insect dung containing plant pollen, begins to show up regularly at the transition between the Silurian and Devonian periods around 420 million years ago. The radiation of the modern angiosperms (the very plant group that is dominant on Earth today) and the modern insect fauna that prey upon them extends from the

Cretaceous period, around 115 million years ago. Although fossil evidence of herbivory is more abundant than evidence for pollination or seed dispersal is, the great species diversity among plants and insects is believed to have been enhanced by both antagonistic and mutualistic coevolution. Indeed, the morphological diversity of flowers, fruits, seeds, and animal pollinators and dispersers observed today provides compelling evidence of a rich and lengthy coevolutionary history.

Sustainability and Ecosystem Management

The sustainability of ecosystems throughout the world depends on an elaborate network of plant-animal interactions that facilitate ecosystem function (energy flow and nutrient cycling). Habitat destruction and the loss of biodiversity, brought about by rapidly expanding human populations and increased resource consumption, is threatening to unravel these core plant-animal interactions to the detriment of natural ecosystems and at great cost to human societies. The remarkable agricultural prosperity of the human race relies profoundly on the persistence of functioning plant-animal interactions. Chief among them is our dependence on insect-pollinated crops and food production for domestic livestock, such as cows, sheep, donkeys, and goats, which provide meat, natural fibers, and labor. Plant-animal interactions are also at the heart of natural processes that threaten human well-being and economic stability, such as the long history of cataclysmic crop damage by insect pests. Consequently, understanding and preserving the coevolutionary relationships between plants and animals is a critical component for a responsible stewardship of natural and agricultural ecosystems on which humans depend.

T. Michael ANDERSON
Wake Forest University

See also in the *Berkshire Encyclopedia of Sustainability* Agricultural Intensification; Agroecology; Biodiversity; Boundary Ecotones; Community Ecology, Complexity Theory; Food Webs; Global Climate Change; Human Ecology; Microbial Ecosystem Processes; Mutualism; Population Dynamics; Refugia

FURTHER READING

Anderson, T. Michael. (2010). Community ecology: Top-down turned upside-down. *Current Biology, 20*(19), R854–R855.

Darwin, Charles. (1859). *On the origin of species by means of natural selection, or the preservation of favoured races in the struggle for life.* London: John Murray.

Darwin, Charles. (1875). *Insectivorous plants.* London: John Murray.

Goheen, Jacob R., & Palmer, Todd M. (2010). Defensive plant-ants stabilize megaherbivore-driven landscape change in an African savanna. *Current Biology, 20*(19), 1768–1772.

Herrera, Carlos M., & Pellmyr, Olle. (Eds.). (2002). *Plant-animal interactions: An evolutionary approach.* Oxford, UK: Wiley-Blackwell.

McNaughton, S. J. (1976). Serengeti migratory wildebeest: Facilitation of energy flow by grazing. *Science, 191*(4222), 92–94.

Owen-Smith, R. Norman. (1992). *Megaherbivores: The influence of very large body size on ecology.* Cambridge, UK: Cambridge University Press.

Pellmyr, Olle; Leebens-Mack, James; & Huth, Chad J. (1996). Nonmutualistic yucca moths and their evolutionary consequences. *Nature, 380*, 155–156.

Pellmyr, Olle; Thompson, John N.; Brown, Jonathan M.; & Harrison, Richard G. (1996). Evolution of pollination and mutualism in the yucca moth lineage. *American Naturalist, 148*, 827–847.

van Dam, Nicole M. (2009). Belowground herbivory and plant defenses. *Annual Review of Ecology, Evolution and Systematics, 40*(1), 373–391.

Willmer, Pat G., & Stone, Graham N. (1997). How aggressive ant-guards assist seed-set in Acacia flowers. *Nature, 388*, 165–167.

Recreation, Outdoor

Earth's natural resources offer humans vast opportunities for outdoor recreation: hiking, swimming, fishing, hunting, camping, and more. In many countries, these activities are dependent on public lands that have been set aside for these uses. Outdoor recreation can impact fragile natural resources through soil compaction and erosion, trampling vegetation, and disturbing wildlife. The future of outdoor recreation is dependent on the conservation of natural resources.

People have been participating in various forms of outdoor recreation for thousands of years. For example, records of the first Olympics date back to 776 BCE, when representatives from the states of the Greek Empire gathered to participate in a running competition. But widespread interest and participation in outdoor recreation in the United States and other Western countries is a relatively new phenomenon tied to increased leisure time, mobility, and wealth, which emerged strongly after World War II.

Parks and Public Lands

In the United States, acquisition and protection of public lands set the stage for future growth in outdoor recreation. Well before public sentiment shifted to a stage where parklands were considered desirable, village commons were set aside for a variety of purposes. Though not originally designed for recreation, the Boston Commons, established in 1634, is often cited as America's first park. The late 1800s saw a number of significant developments in the protection of public lands. Large urban parks such as Central Park in New York City, Rock Creek Park in Washington, DC, and Golden Gate Park in San Francisco were incorporated into American cities. At the state level, New York's Adirondack Park was established in 1892 to be "forever wild." At the federal level, Yellowstone was established as the nation's first national park in 1872. The first forest reserves and wildlife refuges were also set aside during this time period. Early examples of parks and related areas in other countries include Royal National Park in Australia (1879) and Banff National Park in Canada (1885).

With these new public lands came a need to define how they should be managed. During the twentieth century, a number of US laws established the importance of outdoor recreation on public lands. Four outdoor recreation–related agencies established during this period—US Forest Service, National Park Service, Bureau of Land Management, and US Fish and Wildlife Service—are major providers of outdoor recreation opportunities today. From its inception, the National Park Service, created in 1916, had a clear mission to provide for "public enjoyment" of the extensive system of national parks it now manages. Recreational functions of the other agencies were not officially recognized until several decades later, however. In 1960, the Multiple Use Sustained Yield Act dictated that outdoor recreation was to be an official use of US Forest Service lands, equal to the other uses of timber, watershed, grazing, and fish and wildlife. A few years later, Congress established similar mandates for the Bureau of Land Management, an agency created in 1946. In 1997, an organic act for the US Fish and Wildlife Service recognized the appropriateness of wildlife-dependent recreational uses at the nation's wildlife refuges.

During the1960s, additional laws provided protections for public lands that created further outdoor recreation opportunities. In 1964, the Wilderness Act provided protections for wilderness areas on federal lands for the "use and enjoyment of the American people." The National Trails System Act of 1968 encouraged development of trails in both urban and remote scenic areas. In the same year, the National Wild and Scenic Rivers Act provided protection for rivers with "remarkable scenic, recreational, geologic, fish and wildlife, historic, cultural or other similar values."

The protection of parks and public lands for outdoor recreation is not unique to the United States. For example, a network of protected lands in Australia provides the public with opportunities for outdoor recreation. Commonwealth reserves are managed by the Director of National Parks, while additional national parks are managed by individual Australian states and territories. Likewise, an extensive system of national parks in Canada are protected for "public understanding, appreciation, and enjoyment" by the government agency Parks Canada. Throughout the world, the International Union for Conservation of Nature has categorized one billion acres as "National Park" lands (IUCN Category II) that are managed for both environmental protection and outdoor recreation.

In other countries, the public is not so dependent upon the availability of public lands for outdoor recreation. For example, "everyman's rights" or "freedom to roam" traditions in Finland and other Scandinavian countries allow for recreational activities on private lands. In Scotland, responsible recreational access to private lands was formally recognized through the Land Reform Act of 2003.

Long-distance trails provide another means of access to outdoor recreation areas and have been established in many countries throughout the world. In the United States, two long distance trails—the Appalachian Trail in the eastern United States and the Pacific Crest Trail in the western part of the country—were designated as the first National Scenic Trails. Well-known throughout the world, the Inca Trail in the Andes mountain range of Peru leads hikers to the ancient ruins of Machu Picchu. In Japan, the Tōkai Nature Trail extends about 1,700 kilometers (over 1,000 miles) through a series of nature preserves established in conjunction with the trail. The Lebanon Mountain Trail covers 440 kilometers (275 miles) and transects numerous towns and villages.

Administration and Funding

In the United States, several events during the latter half of the twentieth century reflect growing national interest in outdoor recreation. One of the most significant was the establishment of the Outdoor Recreation Resources Review Commission (ORRRC) in 1958. The genesis of the commission was a perceived shortage of outdoor recreation opportunities as a result of the rapidly rising demands associated with the post–World War II period.

The US Congress asked the commission to determine the present and future outdoor recreation needs of the American people, to identify recreational resources, and to recommend policies and programs that would help meet outdoor recreation needs. Following one of the commission's recommendations, the Bureau of Outdoor Recreation was created in 1963. This new agency coordinated federal recreation programs, assisted states with recreation programs, conducted and sponsored outdoor recreation research, fostered regional cooperation among states, and engaged in national recreation planning. Additionally the bureau was charged with managing the newly created Land and Water Conservation Fund (LWCF). Subsidized through a portion of sales on outer continental-shelf oil and gas leases, the LWCF was established to provide financial support to federal agencies and state and local governments for outdoor recreation planning and the acquisition and development of outdoor recreation areas. To receive financial support through the LWCF, states are required to prepare plans—known as Statewide Comprehensive Outdoor Recreation Plans (SCORP)— outlining their priorities for recreational facilities, resources, and programs.

In 1985, a second national commission, the President's Commission on Americans Outdoors, again examined the outdoor recreation needs of the American people. The commission identified a number of trends affecting outdoor recreation, including changes in demographics, technology, and transportation. The commission focused on the importance of having recreational opportunities close to home and encouraged local, in addition to national, action on outdoor recreation. Parks in other parts of the world are funded primarily by central governments and by entrance fees.

Outdoor Recreation Activities

With the creation of the ORRRC came the beginning of a series of national surveys to better understand recreation participation in the United States. The original National Recreation Surveys (NRS) were replaced with the National Survey on Recreation and the Environment (NSRE); these have been conducted every five to ten years since 1960. A number of outdoor recreation activities have been consistently popular among the American public, including picnicking, driving for pleasure, sightseeing, walking, and jogging. In recent decades, there has been an increase in the number of people involved in nature study, photography, and observation. An estimated 203 million Americans participated in some type of nature-based outdoor activity by the end of the first decade of the twenty-first century (Cordell 2008).

Another survey that has been tracking national recreation trends since the 1950s is the National Survey of Fishing, Hunting, and Wildlife–Associated Recreation. An analysis of wildlife recreation trends between 1991 and 2006 showed a slight decrease in participation in hunting and fishing, but an increase in wildlife watching. Visitation has also been tracked by the major federal park and outdoor recreation agencies. By the end of 2009, national parks and national forests received over 285 and 206 million recreation visits per year, respectively (NPS 2010; USFS 2010).

While the United States was the first country to track national outdoor recreation trends, other countries (including Canada, Denmark, Norway, Finland, and Sweden) have conducted national outdoor recreation surveys in more recent years. The content of these surveys varies considerably, making cross-national comparisons difficult. Similarly, several countries (including Australia, Germany, Israel, Japan, Poland, Russia, and Spain) have conducted general leisure-related activity surveys of their citizens. These surveys are also variable and not specifically focused on outdoor recreation activities, however. At present, there is a movement to better coordinate recreation surveys among the research community in Europe.

Contemporary Issues and Challenges

The popularity of outdoor recreation has led to a number of contemporary issues and challenges. For example, outdoor recreation can cause impacts to fragile natural resources, including trampling of vegetation, soil compaction and erosion, water pollution, and disturbance of wildlife. The future of outdoor recreation is dependent upon the conservation of parks and related areas, and outdoor recreation must ultimately be sustainable. High levels of outdoor recreation and inappropriate recreation activities can also threaten the quality of outdoor recreation experiences. Some parks and outdoor recreation areas can be too crowded at some times, and this can degrade the quality of park environments and detract from the quality of the outdoor recreation experience. Some outdoor recreation activities can also conflict with one another: examples include motorized and nonmotorized recreation activities, hunting and wildlife watching, fishing and waterskiing, and skiing and snowboarding. These and related issues need increasing management attention.

Laura ANDERSON and Robert MANNING
University of Vermont

See also in the *Berkshire Encyclopedia of Sustainability* Ecotourism; Parks and Preserves—Marine; Parks and Preserves—National; Parks and Preserves—Wilderness Areas; Tourism

FURTHER READING

Cordell, H. Ken, et al. (2004). *Outdoor recreation for 21st century America*. State College, PA: Venture Publishing.

Cordell, H. Ken. (2008). The latest in trends in nature-based outdoor recreation. *Forest History* Today, Spring 2008, 4–10.

Cushman, Grant; Veal, A. J.; & Zuzanek, Juri. (2005). Free time and leisure participation: International perspectives. Oxfordshire, UK: CABI Publishing.

Hammitt, William, & Cole, David. (1998). *Wildland recreation: Ecology and management*. New York: John Wiley & Sons.

Manning, Robert. (1999). *Studies in outdoor recreation: Search and research for satisfaction*. Corvallis: Oregon State University Press.

Manning, Robert. (2007). *Parks and carrying capacity: Commons without tragedy*. Washington, DC: Island Press.

Moore, Roger, & Driver, B. L. (2005). *Introduction to outdoor recreation: Providing and managing natural resource based opportunities*. State College, PA: Venture Publishing.

National Park Service (NPS). (2010). National Park Service public use statistics office. Retrieved July 12, 2010, from http://www.nature.nps.gov/stats/

United States Forest Service (USFS). (2010). National visitor use monitoring program. Retrieved July 12, 2010 from http://www.fs.fed.us/recreation/programs/nvum/

Zinser, Charles. (1995). *Outdoor recreation: United States national parks, forests, and public lands*. New York: John Wiley & Sons.

Resilience

Ecological resilience is the capacity of a system to withstand disturbance without collapsing and shifting into a different regime. Humankind relies upon a consistent production of ecological goods and services. When an ecosystem's resilience is exceeded and the system shifts into a new regime, the system may be less favorable from a human perspective. Understanding and managing for resilience is thus essential for sustainability.

Humankind relies upon the goods and services provided by ecosystems: clean water and air are two such examples. Resilience is a measure of the disturbance that an ecosystem can withstand before shifting into a different ecological regime, which may provide fewer goods and services. Because it is in humankind's interest to maintain ecosystems in regimes that provide vital ecological goods and services, it is critical to understand resilience.

Background

Resilience is the capacity of an ecosystem to withstand disturbance without collapsing and shifting into an alternate regime, or a different type of ecological system organized around different processes and structures. Examples of alternate regimes are a clear, low-nutrient, low-algae, oxygen-rich lake (oligotrophic) or a turbid, high-nutrient, high-algae, oxygen-poor lake (eutrophic); a coral reef dominated by corals or by macroalgae; and a grassland or a woody-plant-dominated shrubland. Resilience is an emergent phenomenon of complex systems, which means that it cannot be deduced from the behavior of the parts of a system. In other words, a detailed understanding of the wolf and elk populations in Yellowstone does not tell us how the ecosystem as a whole

operates or if it is resilient. The behavior of a system cannot be understood by merely adding together what we know about the parts.

Ecologists understand complex systems to be self-organizing and to have inherent uncertainty, nonlinear dynamics, and emergent phenomena. Complex systems are self-organizing because there is no central entity responsible for directing the processes and functions of the ecosystem. An ecosystem arises instead from the nondirected interaction of the parts; complexity arises over time from many simple interactions. An ecosystem is complex because the whole is more than the aggregation of the parts. Nonlinear dynamics occur when small changes have a disproportionately large effect. Phosphorus levels in a lake may steadily rise over time, with no apparent consequence to the lake, for example, until they rise just a bit more and the lake suddenly tips into a new regime, becoming eutrophic and prone to algae blooms. Ecosystems are also complex adaptive systems; the interaction between the parts and the emergent properties of the whole leads to dynamic changes in the system. These changes have consequences for how scientists understand and manage ecosystems. Understanding the resilience of ecosystems is important, in part because humankind relies upon a consistent production of ecological goods and services, such as drinking water, crop pollination, soil renewal and regeneration, abundant marine life to eat, carbon dioxide storage, and so forth. When an ecosystem's resilience is exceeded and the system shifts into a new regime, the system may become less favorable from a human perspective and produce fewer goods and services.

Ecologists have developed resilience theory since the 1970s to explain the nonlinear dynamics of complex adaptive systems. When the resilience of an ecosystem has been exceeded, the system discernibly changes, such

Figure 1. Resilience

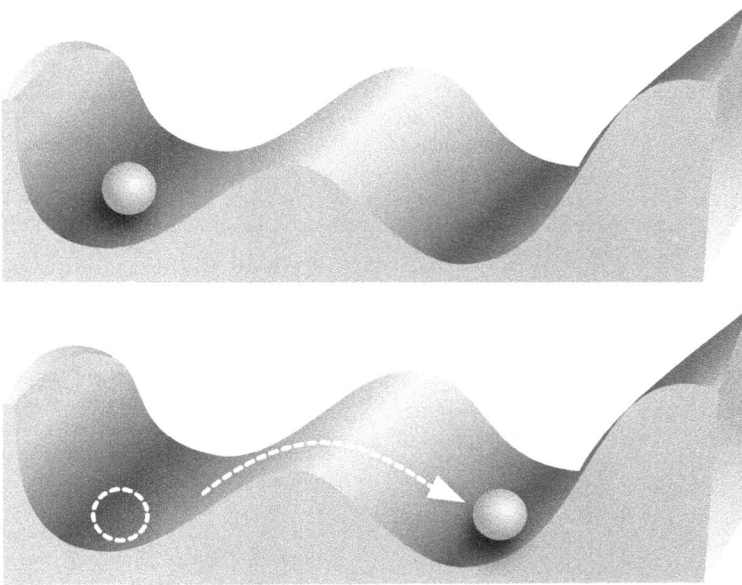

Source: US Environmental Protection Agency.

The figure depicts a conceptual diagram of the basins of attraction for two possible ecosystem states. The position of the ball in the left basin of the upper diagram represents the current state of the system. Ecologists measure the resilience of the system as the amount of disturbance required to push the system from one basin of attraction to another (bottom diagram). The two basins represent two possible alternative stable states, characterized by two different regimes.

as when a lake shifts from a clear to a turbid state. (See figure 1.) The nonlinear dynamics of complex systems make it difficult to predict when that shift might occur, though advances have been made in this area. Small changes to a system may have disproportionately large consequences, and vice versa.

Ecological resilience should not be confused with engineering resilience, which emphasizes the ability of a system to perform a specific task consistently and predictably and to reestablish performance quickly should a disturbance occur. Applying this type of thinking to the management of ecosystems has been extremely problematic. The harvesting of renewable resources such as trees or fish cannot be treated as an engineered system, with predictable and consistent outputs. Ecosystems do not have an equilibrium state, where opposing forces are in balance, as assumed by an engineering definition. An ecosystem exists within a regime. Within a particular regime, the abundance and composition of the species that constitute that regime may change quite dynamically over time. Engineering resilience assumes that ecological

systems are characterized by a single equilibrium state. This assumption is inappropriate for complex adaptive systems such as ecological systems.

Scale is a critical concept to understand when discussing complex systems. Scale typically refers to the spatial extent and temporal frequency of the object or process of interest. In ecosystems, different processes dominate at different spatial and temporal scales. Small and fast processes such as the turnover of leaves on trees are orders of magnitude different from the large and slow processes, such as climate, that drive the location of boreal forest on a continent. Because all these processes occur at discrete spatial and temporal scales, there are thresholds between scales of structuring processes, which are called discontinuities. Neither processes nor structure in complex systems are continuous. Processes operate over domains of scale. These processes are separated by abrupt thresholds that are the transition to a new set of structuring processes. This discontinuous structure is critical to the resilience of ecological systems. Ecologists propose that resilience, or the ability of a system to buffer disturbance and stay in the same regime, results in part from the distribution of function within and across the domains of scales of an ecosystem. The relationship between resilience, discontinuities, and the distribution of function within and across scale has led to formal propositions that allow for quantifiable measures of the relative resilience of different ecosystems.

One such proposition is that the resilience of ecological processes, and therefore of ecosystems, depends partly upon the distribution of function within and across scales. Many ecological functions, such as pollination or seed dispersal, are provided by species. If species that are members of the same functional group operate at different scales, they provide mutual reinforcement that contributes to the resilience of a function, while at the same time minimizing competition among species within the same functional group. Seed dispersal, for example, is an important function that occurs at multiple scales, ranging from the very small in the form of ant dispersal of spring ephemerals to the very large in the form of large vertebrates such as tapirs. Scientists believe resilience is enhanced by the diversity of functions present within a scale and the redundancy of functions distributed across multiple scales. Note that resilience is not driven by the identity of any given element of the system, but rather by the functions those elements provide and their distribution within and across scales.

As of 2011, ecologists detect discontinuities in ecosystems by analyzing animal body-mass distributions. Direct evidence for discontinuities in ecological structure is mounting. Scientists have found discontinuities in animal body-mass distributions in virtually every ecological system they have assessed. They have uncovered patterns that correlate a species location within their particular domain of scale (based on log body mass) to biological phenomena such as invasion, extinction, high population variability, migration, and nomadism. The clustering of these phenomena at the thresholds between domains of scale suggests that variability in resource distribution or availability is greatest at these locations (i.e., discontinuities). These observations support the contention that ecological communities structured by self-organizing dynamics will tend to maintain a similar pattern of discontinuities in animal body-mass distributions despite changes in species composition, at least as long as the processes structuring the system are unchanged.

Origins

The Canadian ecologist C. S. Holling first proposed the concept of ecological resilience in 1973. He recognized

that systems perturbed beyond their capacity to recover could shift into an alternative state or regime. Holling preferred the term *regime* because it emphasizes the controlling processes of a given state of a system. The emphasis on alternative regimes was at odds with prevailing ecological theory of the time, which considered the relevant measure to be return time following perturbation (i.e., engineering resilience). The emphasis on return time was based on the premise that most systems can exist in only one stable state. Holling provided an overview of the origins of the concept in his memoirs:

> Up to that time, a concentration on a single equilibrium and assumptions of global stability had made ecology, as well as economics, focus on near equilibrium behavior, and on fixed carrying capacity with a goal of minimizing variability. Command and control was the policy for managing fish, fowl, trees, herds. . . .
>
> The multi-stable state reality, in contrast, opened an entirely different direction that focused on behavior far from equilibrium and on stability boundaries. High variability, not low variability, became an attribute necessary to maintain existence and learning. Surprise and inherent unpredictability was the inevitable consequence for ecological systems. (Holling 2006)

One of the earliest responses to Holling's original publication was one that refuted it. Wayne Sousa and Joseph H. Connell, both US ecologists, searched the ecological literature to determine if there was evidence for the existence of multistable states in nature. They analyzed published data on time series of population changes of organisms to determine if there was any indication of multiple stable states and found no supporting evidence. This reinforced the single-equilibrium paradigm that dominated in population ecology, and the concept of resilience became, in effect, scientifically dormant.

Over time, as ecologists had access to longer-duration data sets and developed the notion of modeling system behavior (as opposed to the behavior of a part), Holling's concept of resilience began to reemerge. In the late 1990s a grant from the MacArthur Foundation helped create what is currently known as the Resilience Alliance (2011): a "global consortium of institutions that seeks novel ways to integrate science and policy in order to discover foundations for sustainability." The alliance includes universities, government, and nongovernment agencies as partners in a program of research and communications with the goal of integrated social, economic, and ecological sustainability. Since the Resilience Alliance created a global research network focused on developing theory and understanding case studies, publications focused on resilience in social-ecological systems have grown exponentially. Holling's concept of ecological resilience has overshadowed competing definitions focused on return time.

Impact and Application

A premise of any complex system is that surprise and uncertainty are inherent to the system. Conventional ecosystem management has been slow to address surprise and uncertainty in system behavior and has, to a large extent, struggled over the long term to ensure the consistent delivery of ecological goods and services. On the other hand, scientists' ability to replicate a complex system within a lab constrains experimentation on ecosystems. One outcome of the development of resilience theory has been the recognition that we need ecosystem management frameworks that explicitly incorporate planning for and managing uncertainty in ecological systems, as well as for emergent phenomena such as resilience. Ecologists developed adaptive management as a way to conduct safe-to-fail experiments for ecosystems and a way to allow management to occur in the face of uncertainty. Managing for resilience therefore consists of actively maintaining a diversity of functions within and across scales, accounting for thresholds and the nonlinear dynamics that occur at thresholds, and implementing adaptive management and governance. Managing for resilience requires an improved understanding of system-level behavior, in addition to specific, detailed knowledge of parts of the system. Systems in undesirable states can also be highly resilient, however. In such cases the manager's goal is to reduce the resilience of the system and help shift the system to a more desirable regime.

The following propositions constitute the core of managing for resilience:

- Identify the conditions that indicate loss of resilience for the particular system. Recent research demonstrates that there are system-specific conditions that indicate a system is losing resilience and approaching a regime shift. These indicators are measurable and will differ between ecosystems.
- Identify and maintain a diversity of system elements and feedbacks that help keep a system within a desired regime. Maintain the distribution of ecological functions within and across scales that contribute to system resilience.
- Use adaptive management and governance, which are critical to managing for resilience. They treat policy and management options as hypotheses to be put at risk, and thus enhance learning and reduce uncertainty.
- Take management actions to reduce the likelihood of shifts into a different regime. Control invasive species, for example, or monitor and maintain important structuring processes (e.g., fire, hydrology).

Outlook

The success of concepts such as resilience brings with them threats of overuse and misinterpretation. These pitfalls arguably have befallen the concepts of adaptive management and sustainability. Close adherence to the scientific definition and advances in our understanding of regime shifts (and therefore resilience) can prevent this. A challenge to resilience theory is that it is very easy to recognize a system that has undergone a regime shift, but very difficult to recognize when the resilience of a system has been compromised.

Quantification of the resilience of systems is in its infancy and remains poorly developed. Ecologists in the twenty-first century have made advances in detecting early warning of impending regime shifts, usually by focusing on rising variance in key parameters of the ecological system in question. Scientists need to develop leading indicators to manage for resilience, and therefore to develop sound environmental management.

Craig R. ALLEN
*US Geological Survey, Nebraska Cooperative Fish &
Wildlife Research Unit, University of Nebraska*

Ahjond S. GARMESTANI
*US Environmental Protection Agency, National Risk
Management Research Laboratory*

Shana M. SUNDSTROM
*Nebraska Cooperative Fish & Wildlife Research Unit,
University of Nebraska*

Note: The Nebraska Cooperative Fish and Wildlife Research Unit is jointly supported by a cooperative agreement between the US Geological Survey, the Nebraska Game and Parks Commission, the University of Nebraska–Lincoln, the United States Fish and Wildlife Service, and the Wildlife Management Institute. Any use of trade names is for descriptive purposes only and does not imply endorsement by the US government.

See also in the *Berkshire Encyclopedia of Sustainability* Biodiversity; Complexity Theory; Community Ecology; Disturbance; Eutrophication; Extreme Episodic Events; Fire Management; Fisheries Management; Food Webs; Forest Management; Global Climate Change; Hydrology; Groundwater Management; Landscape Planning, Large-Scale; Mutualism; Natural Capital; Regime Shifts; Shifting Baselines Syndrome; Succession

FURTHER READINGS

Allen, Craig R.; Forys, Elizabeth A.; & Holling, C. S. (1999). Body mass patterns predict invasions and extinctions in transforming landscapes. *Ecosystems, 2*(2), 114–121.

Allen, Craig R.; Gunderson, Lance H.; & Johnson, A. R. (2005). The use of discontinuities and functional groups to assess relative resilience in complex systems. *Ecosystems, 8*(8), 958–966.

Biggs, Harry C., & Rogers, Kevin H. (2003). An adaptive system to link science, monitoring, and management in practice. In Johan T. du Toit, Kevin H. Rogers & Harry C. Biggs (Eds.), *The Kruger experience: Ecology and management of savanna heterogeneity.* Washington, DC: Island Press.

Biggs, Reinette; Carpenter, Stephen R.; & Brock, William A. (2009). Turning back from the brink: Detecting an impending regime shift in time to avert it. *Proceedings of the National Academy of Science of the United States, 106*(3), 826–831.

Carpenter, Stephen R., et al. (2011). Early warnings of regime shifts: A whole-ecosystem experiment. *Science, 332*(6033), 1079–1082.

Chapin, F. Stuart, III; Kofinas, Gary P.; & Folke, Carl. (2009). *Principles of ecosystem stewardship: Resilience-based natural resource management in a changing world.* New York: Springer Verlag.

Folke, Carl; Hahn, Thomas; Olsson, Per; & Norberg, Jon. (2005). Adaptive governance of social-ecological systems. *Annual Review of Environment and Resources, 30*(1), 441–473.

Forys, Elizabeth A., & Allen, Craig R. (2002). Functional group change within and across scales following invasions and extinctions in the Everglades ecosystem. *Ecosystems, 5*(4), 339–347.

Garmestani, Ahjond S.; Allen, Craig R.; & Cabezas, Heriberto. (2009). Panarchy, adaptive management and governance: Policy options for building resilience. *Nebraska Law Review, 87*(4), 1036–1054.

Garmestani, Ahjond S.; Allen, Craig R.; & Gunderson, Lance. (2009). Panarchy: Discontinuities reveal similarities in the dynamic system structure of ecological and social systems. *Ecology and Society, 14*(1), 15. Retrieved December 16, 2011, from http://www.ecologyandsociety.org/vol14/iss1/art15/

Gunderson, Lance H., & Holling, C. S. (Eds.). (2002). *Panarchy: Understanding transformations in human and natural systems.* Washington, DC: Island Press.

Gunderson, Lance H.; Allen, Craig R.; & Holling, C. S. (Eds.). (2010). *Foundations of ecological resilience.* New York: Island Press.

Holling, C. S. (1973). Resilience and stability of ecological systems. *Annual Review of Ecology and Systematics, 4,* 1–23.

Holling, C. S. (1992). Cross-scale morphology, geometry, and dynamics of ecosystems. *Ecological Monographs, 62*(4), 447–502.

Holling, C. S. (2001). Understanding the complexity of economic, ecological, and social systems. *Ecosystems, 4*(5), 390–405.

Holling, C. S. (2006). Memoirs. Retrieved October 30, 2011, from http://www.resalliance.org/index.php/holling_memoir

Holling, C. S. (Ed.). (1978). *Adaptive environmental assessment and management.* New York: John Wiley.

Peterson, Garry; Allen, Craig R.; & Holling, C. S. (1998). Ecological resilience, biodiversity, and scale. *Ecosystems, 1*(1), 6–18.

Resilience Alliance. (2011). Homepage. Retrieved December 16, 2011, from http://www.resalliance.org/

Sousa, Wayne P., & Connell, Joseph H. (1985). Further comments on the evidence for multiple stable points in natural communities. *American Naturalist, 125*(4), 612–615.

Walker, Brian H., & Salt, David. (2006). *Resilience thinking: Sustaining ecosystems and people in a changing world.* Washington, DC: Island Press.

Zellmer, Sandi, & Gunderson, Lance H. (2009). Why resilience may not always be a good thing: Lessons in ecosystem restoration from Glen Canyon and the Everglades. *Nebraska Law Review, 87*(4), 893–949.

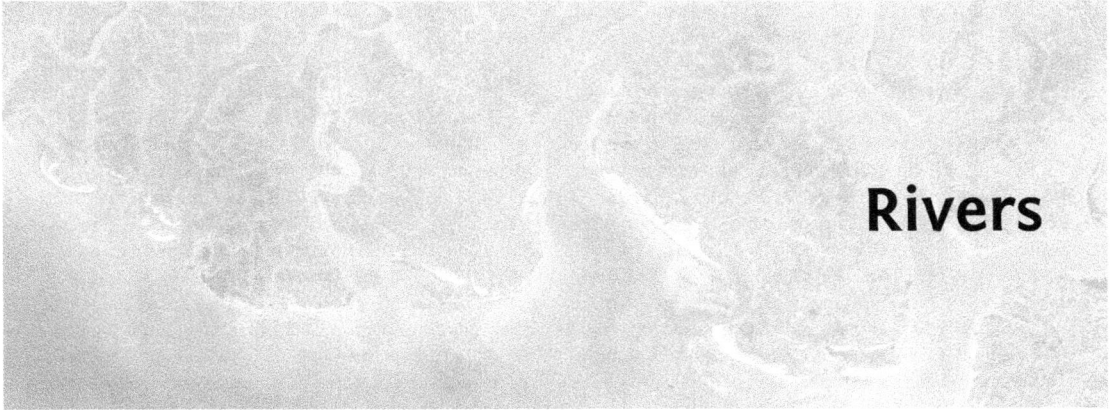

Rivers

The geographer Lewis Mumford's observation—that all great historic cultures thrived by traveling along the natural highway of a great river—is particularly resonant today, as the world's rivers bear the brunt of human manipulation. Pollution and habitat loss (two main side effects of hydraulic engineering), as well as climate change, pose unprecedented challenges to agriculture, manufacturing, urban water supplies, and wildlife conservation.

By common definition, a *river* refers to water flowing within the confines of a channel (as suggested by its Latin root, *ripa*, meaning "bank"). More precisely, a river forms the stem of a drainage system that transports water, soil, rocks, minerals, and nutrient-rich debris from higher to lower elevations. In a broader sense, rivers are part of the global water cycle: they collect precipitation (snow, sleet, hail, rain) and transport it back to lakes and oceans, where evaporation and cloud formation begin anew. Energized by gravity and sunlight, rivers sculpt the land around them, wearing down mountains, grinding rocks, and carving floodplains through the Earth's crust. As carriers of water and nutrients, rivers also provide complex biological niches for fish, sponges, insects, birds, trees, and many other organisms.

Rivers serve as an important source of freshwater for humans as well, not only as drinking water for their own replenishment, but also for their domestic animals and livestock. Hunters find riverine landscapes ideal for locating beaver, ducks, deer, antelope, elephants, lions, and hundreds of other small- and large-game species. Gatherers seek out rivers for nutritional plants and medicinal herbs. The rich soil of the floodplains, meanwhile, nurtures agriculture and horticulture, all the more so when irrigation techniques are utilized to divert and impound water from river channels. The river itself is home to many other natural resources, including fish and mollusks, that provide reliable food supplies, and its channel often acts a conduit of trade that links far-flung peoples together in a web of commerce.

Rivers can be compared by basin size, discharge rate, and channel length, although there are no universally recognized statistics for any of these dimensions. The Amazon forms the world's largest drainage basin at approximately 7 million square kilometers, followed distantly by the Congo (3.7 million square kilometers) and the Mississippi-Missouri (3.2 million square kilometers). As regards discharge rate, the Amazon once again stands supreme at around 180,000 cubic meters per second, followed by the Congo (41,000 cubic meters per second), the Ganges-Brahmaputra (38,000 cubic meters per second), and the Yangzi (Chang) (35,000 cubic meters per second). At about 6,650 kilometers, the Nile is the world's longest river, followed closely by the Amazon (6,300 kilometers) and the Mississippi-Missouri (6,100 kilometers). Other rivers with large area-discharge-length combinations include the Ob'-Irtysh, Paraná, Yenisey, Lena, Niger, Amur, Mackenzie, Volga, Zambezi, Indus, Tigris-Euphrates, Nelson, Huang (Yellow), Murray-Darling, and the Mekong. Two rivers—the Huang and Ganges-Brahmaputra—also stand out for their huge annual sediment load, which makes them especially prone to severe flooding. There is no agreed-upon minimum size, length, or volume for a river, but small rivers are usually called streams, brooks, creeks, or rivulets. Great or small, rivers that form part of a larger drainage system are known as tributaries, branches, or feeder streams.

Although rivers vary greatly in size, shape, and volume, most rivers share certain characteristics. A typical river has its headwaters in a mountainous or hilly region, where it is nurtured by glaciers, melting snow, lakes, springs, or rain. Near the headwaters, swift currents prevail and waterfalls are common, owing to the rapid drop in elevation or the narrowness of the valley though which the river flows. As a river leaves the high region, its velocity typically slackens and its channel begins to meander, bifurcate, or braid. Often its floodplain broadens as it picks up tributary waters. As the river reaches its mouth, it usually loses most of its gradient. It becomes sluggish, allowing some of its sediment load to settle to the bottom, clogging the channel. The river responds by fanning around the sediment deposits, classically forming the shape of a delta (the fourth letter in the Greek alphabet, Δ), before emptying in a lake, sea, or ocean.

Unique climatic and geographic conditions determine a river's annual discharge regime (its seasonal variations in water quantity), but as a rule rain-fed tropical rivers flow more steadily year-round than do snow-fed temperate rivers. A river is called *perennial* if it carries water all or almost all of the time, and *intermittent* or *ephemeral* if it does not. A water-carved channel in an arid region is often called an *arroyo* or *dry gulch* if it carries water only on rare occasions. For hydrologists, the term *flood* refers to a river's annual peak discharge period, whether or not it inundates the surrounding landscape. In common parlance, a flood is synonymous with the act of a river overflowing its banks. In 1887, a massive flood on the Huang caused the death of nearly 1 million Chinese people. In 1988, flooding on the Ganges-Brahmaputra temporarily displaced more than 20 million in Bangladesh. In 2010, flooding on the Indus in the Pakistani provinces of Khyber Pakhtunkhwa, Punjab, and Sindh caused more than 1,500 deaths and left an estimated 6 million people homeless. And in 2010–2011, a series of severe floods in Queensland, Australia, forced more than 200,000 people to evacuate more than seventy towns and cities on the Fitzroy, Burnett, Condamine, Balonne, and Mary rivers.

Human Manipulation of Rivers

Rivers contain only a minuscule portion of the total water on Earth at any given time, but along with lakes, aquifers, and springs they are the principal sources of freshwater for humans as well as for many plants and animals. Rivers are therefore closely associated with the emergence of settled agriculture, irrigated crops, and early urban life. The great civilizations of Mesopotamia (literally "the land between the rivers") from Sumer to Babylonia emerged beginning around 4500 BCE along the Tigris and Euphrates floodplains of modern-day Iraq. Egypt, as the Greek historian Herodotus once famously noted, was "the gift of the Nile." The Huang spawned early Chinese civilization, just as the Indus produced the first cultures of southwest Asia, and the river valleys of coastal Peru shaped urban life in the Andes. "All the

great historic cultures," the geographer Lewis Mumford noted with only slight exaggeration, "have thriven through the movement of men and institutions and inventions and goods along the natural highway of a great river" (McCully 1996, 9).

For most of human history, river manipulation was slight, consisting mostly of diverting or impounding a portion of a river's water for the purpose of irrigating crops. Even modest modifications of this type, however, can have severe environmental consequences. In arid regions, salinization is a common problem. Unless properly drained, irrigated fields slowly accumulate the minuscule amounts of dissolved salts naturally found in soil and water. Over time this salt buildup will eventually render the fields incapable of growing most crop species. Siltation is another common problem, one that often appears when farmers and pastoralists deforest or overgraze river valleys, inadvertently setting in motion excessive erosion downstream. As the silt settles to the channel bottom, it elevates the river above the landscape, making it more prone to flooding.

Ancient Roman, Muslim, and Chinese engineers possessed a sophisticated understanding of the art of hydraulics, as the still-extant aqueducts, canals, and waterworks of Rome, Baghdad, Beijing, and other Eurasian cities amply demonstrate. But river engineering as a mathematical science first emerged in Europe between 1500 and 1800 CE. The crucial breakthrough came when Italian engineers calculated the formula for determining the amount of water flowing in a river at any given time by measuring width, depth, and flow rate. Thereafter, water experts knew how to "tame" a river—that is, manipulate its banks, bed, and velocity in order to contain floods, reclaim land, and promote navigation—with a greater degree of precision, and therefore a higher chance of success, than had previously been feasible.

Today's methods for controlling rivers are remarkably similar to those employed in the past—chiefly the construction of dams and weirs, the reinforcement of banks, and the straightening (and often widening) of channels—but materials and techniques have improved greatly over the past two centuries. Modern dams are designed to store water, regulate minimum channel depth (usually in connection with a lock), generate electricity, or perform all three tasks. Reinforced banks help keep the water in a designated channel, thereby reducing the frequency of flooding and opening up former floodplains for agricultural, urban, industrial, and other human uses. Channel straightening gives a river a steeper gradient and thus a faster discharge rate; by reducing the total length of a river, it also facilitates the transportation of goods between ports. Collectively, these engineering methods transform a mercurial and free-flowing stream (a "floodplain river") into a predictable deliverer of energy, goods, and water (a "reservoir river"). Nowadays the Nile and Yangzi produce kilowatts for industries and cities, just as the Mississippi and Rhine transport freight for companies and consumers, and the Colorado and Rio Grande deliver water for farmers and homeowners.

Environmental Consequences of Hydraulic Engineering

River engineering fosters bank-side economic growth by opening up arable land, reducing floods, promoting trade, and generating electricity, but it also has a disruptive impact on riverine environments. The problems can be divided into two interrelated types: those that compromise the purity of the water in the channel (water pollution) and those that reduce the amount of living space in its channel and floodplain (habitat loss). Both typically result in a reduction in a river's biodiversity.

Water Pollution

Water pollutants can be divided into three broad categories: nutrient based, chemical, and thermal. The most common nutrient-based pollutants are fecal matter from untreated human sewage and agricultural run-off from phosphorus and nitrogen fertilizers. When introduced into a lake or river, these organic substances serve as food for phytoplankton (free-floating algae), which are huge consumers of the water's dissolved oxygen. If the river moves slowly, and if the "algal blooms" are large or frequent enough, the river will gradually become eutrophified (oxygen depleted), with negative consequences for other organisms that require dissolved oxygen for respiration. The Po and Ganges are examples of rivers polluted with sewage and fertilizer.

The most pernicious chemical pollutants include heavy metals (zinc, copper, chromium, lead, cadmium, mercury, and arsenic), and chlorinated hydrocarbons such as polychlorinated biphenyls (PCBs) and dichlorodiphenyl-trichloroethane (DDT). These substances bioaccumulate, that is, they pass unmetabolized from simple organisms to more complex organisms, magnifying in concentration as they move up the food chain. The Mersey, Rhine, Hudson, Ohio, and Donets are all examples of rivers affected by industrial and chemical pollution.

Thermal pollution is a problem on rivers that have numerous nuclear-, coal-, or oil-generated power plants on their banks. The heated wastewater from the plant-cooling facilities artificially raises the water temperature, and the higher temperature in turn affects the type of species capable of living in the streambed. The Rhône and Rhine are examples of rivers afflicted with thermal pollution.

Most of the world's manipulated rivers can loosely be labeled "agricultural" since the majority of engineering projects are geared toward land reclamation (flood control) and the lion's share of dam water is still utilized for the purpose of irrigating crops. As industrialization spreads globally, however, chemical pollutants are increasingly becoming the single greatest threat to river systems; indeed, few rivers remain completely free of any trace of industrial contaminants. Today the cleanliness or filth of a river is often more determined by the average income of the humans who live on its banks than by the number of farms and factories in its watershed. Wealthy nations have invested in urban and industrial sanitation plants over the past fifty years, and water quality has correspondingly improved. Poorer countries, unable to afford these techno-fixes, have seen their rivers continue to deteriorate.

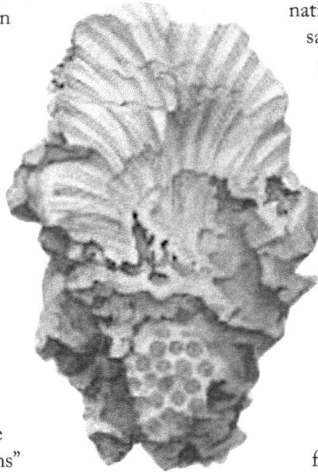

Habitat Loss

Water pollution compromises a river's biological robustness by killing off organisms and by creating an unfavorable environment for nourishment and reproduction, but it is the engineering projects themselves that account for most of the habitat loss on rivers and are thus primarily responsible for the drop in biodiversity. Natural ("untamed") rivers contain an abundance of diverse biological niches: headwaters and tributaries, main and secondary channels, deep pools and islands, banks and bed, and marshes and backwaters. Channels provide longitudinal passageways along which organisms travel, while river edges provide access routes to the adjacent marshes and floodplains where many organisms find their nourishment and reproduction sites. Floodplains nurture trees, shrubs, and reeds, which help stabilize the channel bank while providing shade and protection to other organisms. A river basin hosts a complex web of life, ranging from simple organisms such as fungi, bacteria, algae, and protozoa, to more complex organisms such as flatworms, roundworms, and rotifers ("wheel animals'), and on up to mollusks, sponges, insects, fish, birds, and mammals.

Engineering alters a basin's natural structure in ways that are detrimental to many species. Dams and weirs block a river's longitudinal passageways, making it difficult for organisms to take full advantage of the channel's living space. Migratory fish are particularly hard hit because their life cycles require them to move from headwaters to delta and back again. Most famously, salmon disappeared from the Columbia, Rhine, and many other rivers when they were dammed. Reinforced banks have a similar impact: they sever the links between a river's channel and its floodplain, depriving many organisms of their feeding and breeding sites. As a river loses all or part of its natural channel, bed, banks, islands, backwaters, marshes, and floodplain, it is transformed into a narrow and uniform biological site rather than a broad and diverse one. Typically this results in a precipitous drop in both the number and type of species that it supports.

Aside from reducing the total amount of natural living space on a river, engineering can also trigger dramatic population upsurges in certain species, creating an ecological imbalance. The zebra mussel—a hearty algae eater and rapid reproducer—has migrated from its home in the Caspian Sea to the industrial rivers of the United States and Europe, displacing local mollusk species along the way. Similarly, after the completion of the Aswan Dam in the mid-1930s, snails infected with deadly schistosomes (parasitic worms) began to colonize the Nile's new irrigation canals, debilitating and killing Egyptian farmers and fishermen.

Responding to environmentalists and to reformers from within their own ranks (such as Gilbert F. White), engineers have developed new and more sophisticated methods of river manipulation over the past thirty years. More attention is now paid to preserving the original river corridor as channel beds and banks are fortified and dredged. Dams and weirs are fitted (or retrofitted) with fish ladders to ease fish migration. More floodplain is left intact. In some cases rivers have even been re-meandered and rebraided so that they better replicate the natural conditions that once prevailed on their banks. Nevertheless, the controversial 2007 Three Gorges Dam project on the Yangzi—the largest dam-building project of all time—serves as a reminder that the environmentally unfriendly practices of the past are still in widespread use today.

Most climate scientists predict that global warming will have far-reaching impacts on river systems worldwide. High mountains such as the Alps and Himalayas may begin to shed their snowpack earlier each spring.

Higher evaporation rates may cause some regions to experience significant alterations in their annual precipitation patterns. Warmer water temperatures may make some rivers inhospitable to salmon and other cold-water fish. Rising sea levels may partially or wholly inundate the Netherlands, Bangladesh, and other delta regions. Although the effects on rivers will vary from region to region, collectively these changes will pose unprecedented challenges to agriculture, manufacturing, urban water supplies, and wildlife conservation.

Mark CIOC
University of California, Santa Cruz

See also in the *Berkshire Encyclopedia of Sustainability* Aquifers; Dams and Reservoirs; Glaciers; Oceans and Seas; Waste Management; Water (Overview); Water Energy; Wetlands

FURTHER READING

Cowx, Ian G., & Welcomme, Robin L. (Eds.). (1998). *Rehabilitation of rivers for fish: A study undertaken by the European Inland Fisheries Advisory Commission of FAO.* Oxford, UK: Fishing News Books.

Czaya, Eberhard. (1983). *Rivers of the world.* Cambridge, UK: Cambridge University Press.

Giller, Paul S., & Malmqvist, Björn. (1998). *The biology of streams and rivers.* Oxford, UK: Oxford University Press.

Goubert, Jean-Pierre. (1986). *The conquest of water: The advent of health in the industrial age.* Princeton, NJ: Princeton University Press.

Harper, David M., & Ferguson, Alastair J. D. (Eds.). (1995). *The ecological basis for river management.* Chichester, UK: John Wiley & Sons.

Hillel, Daniel. (1994). *Rivers of Eden.* New York: Oxford University Press.

Mauch, Christof, & Zeller, Thomas. (Eds.). (2008). *Rivers in history: Perspectives on waterways in Europe and North America.* Pittsburgh, PA: University of Pittsburgh Press.

McCully, Patrick. (1996). *Silenced rivers: The ecology and politics of large dams.* London: Zed Books.

Moss, Brian. (1988). *Ecology of freshwaters: Man and medium.* Oxford, UK: Blackwell Scientific Publications.

Nienhuis, Piet H.; Leuven, Rob S. E. W.; & Ragas, A. M. J. (Eds.). (1998). *New concepts for sustainable management of river basins.* Leiden, The Netherlands: Backhuys.

Przedwojski, B.; Blazejewski, R.; & Pilarczyk, Krystian W. (1995). *River training techniques: Fundamentals, design and applications.* Rotterdam, The Netherlands: A. A. Balkema.

Rand McNally and Company (1980). *Rand McNally encyclopedia of world rivers.* Chicago: Rand McNally.

Rogers, Jerry R. (Ed.). (2009). *Great rivers history: Proceedings of the history symposium of the world environmental and water resources congress 2009.* Reston, VA: American Society of Civil Engineers (ASCE).

Soil

Soil is an ecosystem in itself, playing a critical role in nitrogen, potassium, and phosphorous cycles, and performing the crucial function of decomposition for other ecosystems. Although soil degradation occurs because of complex natural processes, the human impact on soil throughout history has intensified erosion at an increasingly rapid pace, especially after World War II, with the cultivation of more marginal lands, the use of toxic chemicals, and the affects of pollution and toxic waste.

Soil, the top 1 or 2 meters of the Earth's surface, is tremendously abundant with life. No subject (or substance) is more central to the environment than soil. It is the ultimate ecosystem, combining all the elements of other ecosystems and carrying out the critical ecosystem function of decomposition for most other ecosystems. Befitting this centrality, many disciplines study aspects of soil, but generally students of soil come from the geosciences, agronomy (a branch of agriculture dealing with field-crop production and soil management), and ecology. The branches of soil science (pedology) include genesis and classification, geography, fertility and crop productivity, chemistry, microbiology, physics, and erosion and conservation. Taken together, the disciplines study soil from at least three perspectives: geologically as a part of the Earth's surface, ecologically as an ecosystem, and agronomically as a medium for plant growth.

Soil, as a geological layer and ecosystem, bears the strong imprint of humankind. Indeed, the more scientists analyze soils around the natural world, even in the Amazonian wilderness, the more they find anthrosols (soils bearing the effects of long-term human use). Today humans cause more geomorphic change than any other single Earth surface process (e.g., rivers, wind, and glaciers), and the largest human agent of landscape change is soil erosion.

Soils in the Environment

The pedosphere (soil sphere) derives from and interacts with other Earth "spheres": the lithosphere (the outer part of the solid Earth, composed of rock and consisting of the crust and outermost layer of the mantle), the hydrosphere (the aqueous envelope of the Earth, including bodies of water and vapor in the atmosphere), the atmosphere (the gaseous envelope surrounding the Earth), and the biosphere (the part of the world in which life can exist). Every biogeochemical cycle (a term relating to the partitioning and cycling of chemical elements and compounds between the living and nonliving parts of an ecosystem) has a major sink (a body or process that acts as a storage device or disposal mechanism) and many transformations through soil. Soil's role is paramount in the nitrogen, potassium, and phosphorous cycles, and ecosystem productivity would be greatly decreased without soil nitrogen fixing bacteria. In an era of global warming with carbon dioxide as the chief anthropogenic (human-caused) contributor, it is important to remember, as numerous researchers have concurred, that the world's soil stores more carbon than is circulated in the atmosphere. Land-use changes can accelerate the transfer of soil carbon into atmospheric carbon dioxide and also sequester it from the atmosphere. Therefore, the role of soil in both causing and preventing global change is important and sometimes neglected.

Equally important is the function of soil in water filtration and the hydrologic cycle, the sequence of conditions through which water passes from vapor in the atmosphere through precipitation upon land or water surfaces and ultimately back into the atmosphere as a result of evaporation and transpiration. Water infiltrates soils, where it serves as the main reservoir for plant water use, and water percolates through soils into groundwater.

In this cycle many processes filter and purify water. These properties have led humans to use soils as natural water treatment plants to remove a myriad of contaminants. Histosols—wetland soils such as peat, which are composed primarily of organic materials—perform such ecosystem functions, saving individuals, businesses, and governments worldwide billions of dollars a year. Unfortunately, Histosols and the wetlands in which they form are rapidly disappearing around the world as humans develop land for other uses.

Soil Definitions and Formation

In some sense soil can simply be considered as the layer of mineral and organic aggregates that covers much of the continental surface of the Earth. But most of the large expanses of glaciers, shifting desert sands, and rock pavements don't grow plants and are not really soil. Soil has texture (sand, silt, and clay particles), structure (how these textures adhere together in aggregates called "peds"), organic matter (dead plant and animal material), gases and water in micropores and macropores, and an incredible myriad of life forms that parallel above-ground ecosystems, but with more emphasis on decomposers. These soil ecosystems are still the least understood in nature.

A host of processes—weathering, particle translocation, and organic matter addition or removal over time—act upon the soil medium to produce horizons, or soils zones differentiated horizontally by color, texture, structure, and chemistry. Generally, through a soil profile, from the top down, soils can have O, A, E, B, and C horizons. The O horizon occurs especially in forest and wetland soils and indicates a layer predominantly made up of partly decomposed plant matter. The A horizon in most soils is the uppermost mineral horizon with high amounts of organic matter, structure that enhances aeration and infiltration, and high fertility. The E horizon is a zone of removal, where leaching of nutrient ions and organic matter occurs, whereas a B horizon is a lower zone of addition or transformation of clay and weathered minerals. Finally, a C horizon is the soil's parent material altered by mineral weathering. A soil could have all of these horizons or just one, such as Histosol or peat composed of O horizons, and still be soil.

Scientists analyze and classify soils based on their diverse and/or complicated morphologies (forms and structures) and on their genesis. Soils' constituents are the product of the geomorphic processes (those relating to form or surface features of the Earth), including chemical and biological weathering, and the history and time lapsed since the soil medium formed. In this sense, it is useful to conceptualize soil as a living ecosystem controlled by its factors of formation: $S = f(cl, o, r, p, t, . . .)$, where soil ($S$) is a function ($f$) of climate ($cl$), organisms ($o$), relief ($r$), parent material ($p$), time ($t$), or other possible factors (which replace the dots in the equation). This factorial method considers soil as a part of the whole ecosystem, and it recognizes that soil evolves over time.

The factors of soil formation allow people to study sequences of soils that vary by only one of the five factors, such as relief. For example, a series of soils across a slope is known as a "catena" or "soil toposequence," and soils along it would range from the crest of the slope downward to the shoulder, back, foot, and toe of the slope. The cause for soil variation on this slope is the slope gradient and position, and the results of this would be differences in erosion, deposition, water drainage, and weathering. These topographic and other sequences, together with the horizontally varied layers of soil, give a range of characteristics that groups of soils tend to have. This range provides baselines to estimate change in soils such as sediment burial of topsoils and truncation of topsoils by erosion.

Soil Classification

Classifying soils has an ancient history, and most agricultural societies, including the Romans, Greeks, Aztecs, and Maya, have had soil taxonomies. Farming peoples depend on the soil landscape, and successful farmers differentiate regions where crop yields are good or where they lag after a period or time. The Maya of the Yucatan Peninsula, for example, use folk soil taxonomy based on numerous potential cropping areas. The taxonomy includes seven principal arable land taxa (classes) and many other modifying taxa. These taxa names usually refer to a soil characteristic of topography, texture, or color, but folk soil taxa have significance that transcends mere description. Each taxon is richly meaningful about the crops that work best for a particular soil, especially maize varieties.

Many contemporary soil taxonomies had their origins in the genetic classification of the Russian school of the later nineteenth century led by the so-called father of pedology, Vasily V. Dokuchaev. The underlying idea is that soils are unique bodies formed by the major factors and many pedogenic processes. This idea passed through the soil scientist K. D. Glinka in Germany to C. F. Marbut, the geology professor who introduced the idea in the United States in the 1920s. Two of the classifications used most are those of the United States Department of Agriculture (USDA) and the United Nations Food and Agriculture Organization (FAO), although many nations have their own classifications. The USDA classifies the world's soils in twelve general orders based on a variety of soil characteristics related to the five main soil-forming factors.

Six soil orders, Entisols, Inceptisols, Aridisols, Andisols, Gelisols, and Histosols, have straightforward definitions. Entisols are juvenile soils, whereas Inceptisols are more developed but still immature. Both form in recent deposits such as floodplains, steep, active slopes, or beaches; together they make up more than 25 percent of the Earth's soil surface. Aridisols form in arid environments and make up nearly 12 percent of the Earth's soil surface. They have buildups of salts or carbonates and, where not yet salinized, often salinize quickly with irrigated agriculture. Many of the first hydraulic civilizations (irrigated agriculture–based societies) farmed on Aridisols and Entisols. Andisols are soils formed in recent volcanic materials and cover less than 1 percent of the Earth's soil surface. Gelisols have dark, organic surfaces and form in permafrost; these cover almost 9 percent of the Earth's soil surface, including most of Alaska. Histosols are peat and muck, forming from organic matter; they cover only about 1 percent of the Earth's soil surface.

Defining the remaining soil orders requires a bit more elaboration. Mollisols are the world's most nutrient-rich and organic-matter-rich soils, making up about 7 percent of the Earth's soil surface. They form in some tropical limestone terrains but predominantly in the world's prairies and steppes in the North and South American Midwest and central Eurasia. Because they are so fertile, cultivation covers most of their land area, although urban sprawl is spreading over notable areas of these prime agricultural lands. Alfisols generally form under temperate and mixed deciduous (leaf-dropping) forests and make up almost 10 percent of the Earth's soil surface. These are also fertile soils that have both nutrient and clay enrichment in subsoils but lack the dark, organic topsoils of Mollisols. Ultisols make up about 8 percent of the Earth's soil surface and are similar to Alfisols with their enriched clay subsoils, but more intense and longer weathering makes them more acidic and lower in organic matter and nutrients. Alfisols occur in pockets all around the world, but the redder, more leached Ultisols occur more in subtropical climates such as the southeastern United States, southern China, and India. Mediterranean climates often have "terra rosa" soils, which are usually Alfisols and Ultisols. Spodosols are another acidic forest soil, but they tend to form in sand plains of the subtropics (e.g., Florida) or under coniferous boreal (northern) forests (e.g., Siberia). These are the most colorful of soils, with black organic topsoils that overlie pale, sandy horizons, which in turn overlie red and yellow, iron-rich and aluminum-rich horizons. They make up only about 2.5 percent of the Earth's soil surface, but six US states, including Alaska and Florida, have declared them their state soils. Vertisols make up about 2.5 percent of the Earth's soil surface. They are the most dynamic soils because they are dominated by highly expansive clays that self-plow and shrink and swell with dehydration and hydration. They occur in extensive clayey pockets around the world in India, Central America, Africa, and the United States. Oxisols are usually red, clay-rich soils that are composed of highly weathered iron and aluminum oxides. These cover about 7.5 percent of the Earth's soil surface but only in the tropics or as remnant soils of ancient tropical climates in temperate latitudes. Oxisols predominate in tropical rain forests such as Amazonia, southeast Asia, and central Africa.

Soil Fertility

Soil is a medium of growth for nearly all plants on Earth. Soil fertility is thus vital, and many characteristics contribute to soil fertility, including nutrient availability, lack of toxic elements, texture, structure, and a healthy microbial ecosystem. Plants need six macronutrients in higher quantities from soil: nitrogen, phosphorus, potassium, calcium, magnesium, and sulfur. Plants also need eight micronutrients at smaller quantities: iron, manganese, copper, zinc, boron, molybdenum, chlorine, and nickel. These nutrients come from the weathering of minerals in the soil's parent material, from recycling of other plants, and some from atmospheric deposition, wind-eroded particles, and tephra (solid material ejected into the air during a volcanic eruption).

Within the soil, clays and organic matter are the storehouses of soil nutrients. These store and release nutrients with weathering and mineralization, and they hold nutrient ions at and near their surfaces in solution. Uptake of these nutrients by plants occurs most effectively where soils harbor rich ecosystems. Most important of all are soil fungi called "mycorrhiza," which develop mutualistic relationships with plants, increasing plant uptake of nutrients and requiring only some carbon compounds in return. Plentiful mycorrhiza, for example, allow the important tropical crop of cassava to prosper in soils low in phosphorous.

Some elements and nutrients, such as aluminum or magnesium, can be at levels that are too high. For example, serpentine is a type of rock that can weather out toxic levels of magnesium and produce barrens or spotty plant communities resistant to high levels of this macronutrient.

History of Erosion

The history of human-induced soil erosion can be understood by examining events and occurrences in three great waves. The first wave started about 2000 BCE as humans expanded out of their floodplains in Bronze Age China, the Middle East, and south Asia. The second wave started with the fifteenth-century European expansion around the world, pioneering new lands that Old World farmers had never before encountered. Soil erosion coursed through these colonial and other newly plowed lands because the settlers had developed their practices in the milder climates and gentler slopes of Europe. The third wave started after World War II with the tremendous expansion of agriculture onto marginal lands, especially in the tropics, where intense rain and steep slopes combined to produce some of the world's highest rates of erosion. The three waves were similar in one respect: pioneer farmers moved onto lands that they did not understand and that were usually more erodible than lands with which the farmers were familiar.

Soil Degradation: Outlook for the Future

The causes of soil degradation are numerous and cumulative—and they depend on natural land formations and processes as well as anthropogenic impact. For example, natural soil erosion results from a complex combination of wind and water conditions acting to sculpt the land over different temporal and spatial scales, and yet it can vary due to the soil's inherent resistance to erosion, the amount of vegetative land cover, or the steepness and length of slopes.

Human-induced erosion often occurs much faster and can produce changes that render upland soils unusable for agriculture and downstream floodplains glutted with sediment. One obvious cause of intensifying degradation is intensifying land use: more plowing, more plowed area, more irrigation, generally more cultivation of more marginal lands, and more toxic chemical use and oil exploitation. Greater chemical use, especially nitrogen fertilizer, has led to higher yields, but large areas of polluted soils have also developed, especially from metal smelting, petroleum, and a stew of other toxic wastes. The expansion of these processes occurred especially after World War II and spread to the developing world with heavy industry in the 1970s.

United Nations Agenda 21 (a global partnership adopted at the United Nations Conference on Environment and Development at Rio de Janeiro in 1992) underscored the problems of land degradation and the need for soil conservation. Spurred by the conference proceedings, researchers conducted studies in the early to mid 1990s and estimated that 23 percent of global arable land had been degraded to an extent that limited its productivity. A number of ensuing multilateral agreements on an international level began to include provisions that could be used to promote the sustainable use of soil, although the provisions were generally tangential. Key global instruments relevant to soil include the 1994 United Nations Convention to Combat Desertification (which 194 nations have signed as of 2011), the 1995 Convention on Biological Diversity, the 1995 United Nations Framework Convention on Climate Change and its 1997 Kyoto Protocol (and presumably its successor).

The European Union drafted a proposal for a Directive of the European Parliament and of the Council to establish a framework for the protection of soil in 2006, combining the integration of soil concerns into policies relevant to other sectors. Among other considerations it addresses the prevention of threats to soil (those occurring naturally and those caused by human activity) and the mitigation of their effects based on the identification and remediation of contaminated sites. This proposed directive also considers the preservation of the capacity of soil to perform environmental, economic, social, and cultural functions. And yet making such a framework effective requires the integration of soil protection in the formulation and implementation of national and European Community policies and increasing public awareness of the need to protect soil (European Union 2006). As of 2010 there has been insufficient consensus to move the framework forward.

Soil conservation movements worldwide will depend on the ways in which governments and businesses develop rules, treaties, and laws that can effectively address the natural and human causes of land degradation, such as nutrient removal, salinization, waterlogging, pollution, and land development (urban and suburban sprawl). Essentially each process either removes or limits soils from agricultural production or other ecosystem functions, including soil's ability to sequester carbon dioxide, filter polluted air, provide habitat, and function in the nitrogen, hydrologic, and other biogeochemical cycles. The future of the biosphere greatly depends on how effectively we sustain the Earth's pedosphere, thereby allowing soil to perform its inherent functions.

Timothy BEACH
Georgetown University

See also in the *Berkshire Encyclopedia of Sustainability* Agriculture (*several articles*); Dung; Fertilizers; Green Revolution; Guano; Manure, Animal; Manure, Human; Rivers; Water (Overview); Wetlands

FURTHER READING

Amundson, R., Harden, J., & Singer, M. (Eds.). (1994). *Factors of soil formation: A fiftieth anniversary retrospective* (Special Publication No. 33). Madison, WI: Soil Science Society of America.

Beach, Timothy. (1994). The fate of eroded soil: Sediment sinks and sediment budgets of agrarian landscapes in southern Minnesota, 1851–1988. *Annals of the Association of American Geographers, 84*(1), 5–28.

Beach, Timothy. (1998) Soil catenas, tropical deforestation, and ancient and contemporary soil erosion in the Petén, Guatemala. *Physical Geography, 19*(5), 378–405.

Bennett, H. H., & Chapline, W. R. (1928). Soil erosion: A national menace (USDA Circ. No. 33). Washington, DC: USDA.

Birkland, P. W. (1998). *Soils and geomorphology* (3d ed.). New York: Oxford University Press.

Buol, S. W., Hole, F. D., McCracken, R. J., & Southard, R. J. (1997). *Soil genesis and classification* (4th ed.). Ames: Iowa State University Press.

Butzer, K. (1976). *Early hydraulic civilization in Egypt.* Chicago: University of Chicago Press.

European Union. (2006). Proposal for a directive of the European Parliament and of the Council establishing a framework for the protection of soil and amending Directive 2004/35/EC. Retrieved November 30, 2010, from http://eur-lex.europa.eu/LexUriServ/LexUriServ.do?uri=CELEX:52006PC0232:EN:NOT

German Advisory Council on Global Change. (1994). *World in transition: The threat to soils: Annual report.* Bonn, Germany: Economica Verlag.

Harbough, W. (1993). Twentieth-century tenancy and soil conservation: Some comparisons and questions. In D. Helms & D. Bowers (Eds.), *The history of agriculture and the environment* (pp. 95–119). Berkeley and Los Angeles: University of California Press.

Hillel, D. (1991). *Out of the earth: Civilization and the life of the soil.* Berkeley and Los Angeles: University of California Press.

Hooke, R. L. (2000). On the history of humans as geomorphic agents. *Geology, 28*, 843–846.

Hughs, J. D. (1994). Sustainable agriculture in ancient Egypt. In D. Helms & D. Bowers (Eds.), *The history of agriculture and the environment* (pp. 12–22). Berkeley and Los Angeles: University of California Press.

Jenny, H. (1941). *Factors of soil formation.* New York: McGraw-Hill.

Lowdermilk, W. C. (1953). Conquest of the land through seven thousand years (Agricultural Information. Bulletin No. 99). Washington, DC: USDA.

Mann, C. C. (2002). 1491. *The Atlantic, 289*(3), 41–53.

McNeill, J. (2000). *Something new under the sun: An environmental history of the twentieth-century world.* New York: Norton.

National Research Council. (1993). *Soil and water quality.* Washington, DC: National Academy Press.

Pimentel, David, et al. (1995). Environmental and economic costs of soil erosion and conservation benefits. *Science, 267*, 117–123.

Reich, P., Eswaran, H., & Beinroth, F. (2003). Global dimensions of vulnerability to wind and water erosion. Retrieved February 14, 2003, from http://www.nrcs.usda.gov/technical/worldsoils/land-deg/papers/

Soil Survey Staff. (1998). *Keys to soil taxonomy* (8th ed.). Washington, DC: USDA Natural Resources Conservation Service.

Tanji, K. K. (1990). The nature and extent of agricultural salinity problems. In K. K. Tanji (Ed.), *Agricultural salinity assessment and management* (American Society of Civil Engineers Manuals and Reports on Engineering Practice No. 71) (pp. 1–17). New York: American Society of Civil Engineers.

The twelve soil orders. (n.d.). Retrieved February 14, 2003, from http://soils.ag.uidaho.edu/soilorders/index.htm

United Nations Food and Agriculture Organization (FAO). (1996). *Our land, our future.* Rome: United Nations Food and Agriculture Organization and United Nations Environment Programme.

United Nations Food and Agriculture Organization (FAO). (1998). *World reference base for soil resources.* Rome: United Nations Food and Agriculture Organization and International Society of Soil Science.

Van Andel, T. H. (1998). Paleosols, red sediments, and the old Stone Age in Greece. *Geoarchaeology, 13*(4), 361–390.

Wild, A. (1993). *Soils and the environment: An introduction.* Cambridge, UK: Cambridge University Press.

Yaalon, D. H. (2000, September 21). Why soil—and soil science—matters? Millennium essay. Nature, 407, 301.

Succession

Succession is temporal change in ecosystem structure that can be initiated either naturally or by humans. Ecologists and ecosystem managers use different models to understand and predict this change in order to promote sustainability. The traditional linear models for succession have been challenged by newer nonequilibrium models that emphasize abiotic controls and multiple endpoints.

Ecological succession refers to the change in ecosystems and their constituent plant communities that occurs as the organisms in the ecosystem respond to and modify that ecosystem over time. Succession models are theoretical frameworks for describing and interpreting the development of plant communities and ecosystems, and they have had profound impact on the way that people interpret and manage the land. Assumptions derived from ideas about how ecosystems develop and change also influence how people think about sustainability. The classic succession model posits that soils, plants, and the animals associated with them go through various stages of development until they reach a stable equilibrium with their physical environment. This final stage is termed a climax state, achievable in the absence of disturbance that sets the community or ecosystem back. For many, ecological sustainability has come to mean achieving and maintaining some sort of equilibrial state. More recent theories of ecosystem change, however, cast doubt on the viability of an equilibrial state as a goal or measure of the sustainability of a system. Newer models, based on nonequilibrium theories, offer a more broadly applicable framework for understanding and evaluating the sustainability of managed ecosystems.

Linear Succession Model

The classic model of succession views ecosystem development as a linear process, driven by competition among living organisms. The thinking behind this model essentially began in the early twentieth century and was elaborated in the 1920s by the ecologist Frederick Clements while studying plowed fields in the Midwest. He fully developed the concept of linear, deterministic succession to describe vegetation response to human disturbance. This concept gained such traction that it is now often referred to as the Clementian succession model. In the decades following his work, ecologists have debated the particulars of the model, and whether succession follows from the organic development of ecosystems, the individual characteristics of plants, the facilitation or suppression of new plants by existing plants, or the initial floristic composition on the site.

The standard illustration of primary succession begins with a volcanic eruption and fresh lava that eventually weathers. Plant seeds brought in by wind or animals take root. Soils form, and a predictable succession of plant communities or ecosystem states—from grass and shrubs to forest—ensues. Canopy closure shades out the herbaceous and shrubby pioneer plants, and the forest matures and eventually reaches a stable state of equilibrium when it becomes "old growth." Secondary succession takes place after less traumatic events or disturbances that do not require a new soil to form, such as when a forest is burned or cut down and then recovers, progressing from grass and shrubs back to trees.

The environmental characteristics and species interactions at each site influence the community at the endpoint, whether forest or desert scrub. The collective stages of plant community succession at a site, referred to as *successional seres*, may provide a series of habitats for distinct wildlife species. For example, the spotted owl of California depends on an old-growth forest habitat and is often described as a climax species. Various attempts have been made to correlate maximum biodiversity and other features with climax stages. Some observers and ecosystem managers still evaluate the condition of an ecosystem or plant community based on how close its current state is to the predicted climax, although many ecologists consider this to be a misleading and inaccurate method of assessment.

Succession models based on succession toward an equilibrial state imply that any force that drives an ecosystem away from climax is detrimental. This idea has influenced the way ecologists and ecosystem managers have interpreted landscapes and assessed sustainability. For example, in much of the world it has provided a rationale for suppressing indigenous and traditional patterns of natural-resource management and use. The effect of humans on the ecosystem, seen as a form of disturbance, has often been assumed to be detrimental to the ecosystem's condition because it moves the ecosystem further from a posited climax state. Ecosystem managers and conservationists have treated natural and human-made fire as a detrimental disturbance to ecosystems for decades, and only recently have begun to fully accept managed fire as a shaper of sustainable ecosystems. The anthropologists James Fairhead and Melissa Leach argued in a 1995 article that environmental historians misinterpreted a landscape in Africa because they assumed that human actions were inherently degrading to ecosystems and would cause a loss rather than a gain of forest cover. The use of fire by indigenous people was eventually found, in fact, to contribute to forest renewal and human sustenance. Underlying this misinterpretation is the classic succession model now largely rejected by ecologists but still widely influential among managers.

Alternative Succession Models

An alternative to the classic view, nonequilibrium ecology holds that ecosystem characteristics are more often influenced by disturbance and abiotic (nonliving) factors than by the biotic (organic) interactions that are used to explain the linear pattern of succession. When disturbance is frequent, severe, and unpredictable, stochastic models may be a better fit to the observed ecological changes. Stochastic models consider the outcomes of succession to be unpredictable, though attempts at prediction may be based on probabilities derived from historical outcomes. Most ecosystem dynamics, however, fall somewhere between an undisturbed progression

from one set of species to another, driven by competition and other biotic factors, and a completely unpredictable system that responds ad hoc to unanticipated disturbance. Instead, a model that recognizes the persistence of relatively stable ecosystem configurations and acknowledges that there are multiple possible states and pathways among them has proven useful for understanding ecological dynamics. Such models are termed state and transition models. Table 1 on page 138 compares the three approaches.

For those interested in sustainable management of ecosystems, the simple linear deterministic model of succession is easy to apply when evaluating the potential sustainability of managed systems, but it has low predictive power in many systems. Lands used for agriculture, grazing, timber, or many other types of ecosystem service production are inherently managed for vegetation conditions that would be considered nonclimax, or distanced from the supposed equilibrial state. State and transition models, on the other hand, suggest that management is better understood as a process of choosing among possible stable states and maintaining those states rather than seeking a single equilibrial climax. They accommodate the role of stochastic processes in ecosystem change, as well as the driver of classic succession, plant competition. State and transition models can also evaluate the role of human interaction in ecological change, unlike the linear succession model.

A state and transition model is research intensive. It allows the manager to identify ecosystem states that are stable within management horizons, and to gather data that identifies states and explains shifts from one state to another. Management practices that change or maintain sites can be identified, tested, and incorporated into the model, which makes it amenable to adaptive management. In a forest, for example, fire frequency and intensity, natural or human caused, may maintain the stability of a state or cause a transition among the possible stable states on a site, affecting the characteristics of the forest indefinitely. A linear succession model predicts a single path after fire from "weedy species" like grasses to a climax state of mature trees, while a state and transition model accommodates different stable endpoints or outcomes depending on management practices or natural events. States and transitions are defined by data collected within a site of well-defined environmental characteristics, including soils, climate, slope, and other factors. As more is learned about the ecosystem, states and the transitions among them may be better defined and understood. State and transition models incorporate new information and can be corrected and elaborated as more data is provided. Because they do not rely on a singular pattern of development to explain ecosystem

TABLE 1. Comparison of Three Types of Models for Explaining Species Composition and Ecological Change

Model	System Characteristics	Site Conditions Under Which Model Might Work
Linear deterministic succession (Clementian succession)	• Long- and short-term prediction • Initial conditions are known but have minor influence on predicting succession • Competition and other plant-animal interactions are drivers within given site conditions • A single state commonly develops in the absence of disturbance	• Abiotic conditions are stable • Disturbance is rare and extrinsic to the community • Low level of environmental heterogeneity
Stochastic	• Short-term predictions are more accurate • Initial conditions are often unknown and important predictors • Random events are an important influence on succession • Order of propagule arrival may have strong influence	• Frequent and relatively intense unpredictable disturbance • Small spatial and temporal scales
Alternative stable states (states and transitions)	• Initial conditions known or unknown • Environmental conditions may be important drivers of succession • Random events may be important • The contributions of resilience and thresholds to state stability are important • Disturbance may be important factor, intrinsic or extrinsic to the ecosystem • Abiotic factors may or may not mediate competition	• Adaptable to any spatial and temporal scale or pattern

Source: authors.

change, they can define research needs for understanding the dynamics of an ecosystem. The major disadvantage to state and transition models is the up-front need for more information about states and transitions, but because they are based on data rather than solely on theory, their predictive capacity is high compared to Clementian models in ecosystems where abiotic factors are important drivers of vegetation change. Specific methods for better incorporating reliable information into management practices are now well developed and known as adaptive management.

Ecosystem Stability

While the state and transition model of ecological succession focuses on multiple possible outcomes, ecosystem managers remain concerned with establishing stability, and certain theoretical frameworks are important to finding the keys to sustainability. Understanding the factors that predict the stability of an ecosystem state can

result in management practices that will sustain that state within the bounds of likely climate, social, and economic change. The concept of resilience is used to describe an ecosystem state with strong feedbacks or responses that help maintain the state. For example, openings in a grassland may stimulate more seed production in nearby grasses because increased nutrients are available, so the openings will eventually close again. Feedbacks can also be destabilizing, such as when an invasive grass increases the likelihood of fire in sagebrush steppe, leading to more frequent fire, even more grass, and a state change, or transition, to grassland. Ecosystem managers aim to identify and support feedbacks that promote stability to enhance ecosystem sustainability.

The concept of thresholds is also important in understanding what contributes to stability in the state and transition model. Changing from one ecosystem state to another may involve crossing a threshold that has directionality—it is easier to go in one direction than another. For example, a state and transition model constructed for Australian eucalypt forest predicted that when overharvest completely removes trees, a subsequent influx of saline groundwater shifts the site from a eucalyptus-dominated lowland to a salty flat that can support only salt-tolerant plants. Returning to the eucalypt-dominated state, if feasible, would require major and costly interventions. In other words, the overharvest causes a transition that crosses a threshold. Understanding transitions, resilience, and thresholds can be invaluable when assessing opportunities for restoration and their likelihood of success.

A simplified "ball and cup" diagram can be used to illustrate the ideas of resilience, thresholds, transitions, and stable states. (See figure 1.) The bowl of the cup is a stable state, and in order to shift to another state, the ball must transition over a threshold, represented as the rim of the cup. The ball can be thought of as moving within the cup due to natural environmental variations and

even succession. In the diagram, which illustrates changes in a California oak woodland, oaks may be removed by fire or harvest, a transition with a relatively low threshold. If the oaks are completely removed and there is no resprouting, however, it will take planting and perhaps protection of regrowth to restore the oak woodland, a difficult transition that is unlikely to occur naturally. (See figure 1 below.) It must be noted that these models need validation by empirical tests, and they fit specific sites with particular environmental configurations.

Outlook and Challenges

There are three areas of major challenges and debates concerning new theories of succession or ecosystem change.

Figure 1. Resilience, Thresholds, Transitions, and Stable States

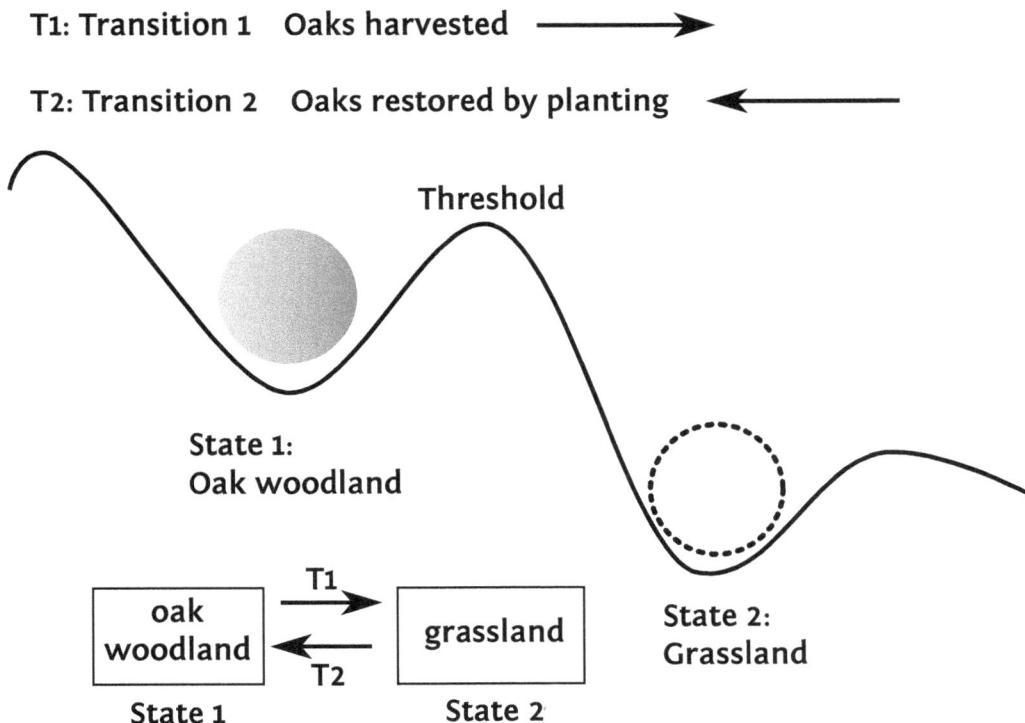

T1: Transition 1   Oaks harvested ⟶

T2: Transition 2   Oaks restored by planting ⟵

Source: authors.

This "ball and cup" model illustrates the role of thresholds and resilience in transitions among states. Resilience operates like gravity to keep the ball in the cup. Thresholds and resilience support the stability of a state, keeping it within recognizable bounds. The ball represents the ecosystem status, and the depth of the cup illustrates the state's resilience to disturbance.

First, many managers prefer the linear succession model because it can greatly simplify management planning and monitoring and provides an implicit goal. The models work fairly well in many ecosystems, such as temperate hardwood forests. Even though their use has become a professional norm, however, linear succession models are open to criticism as premised on a way of looking at the world that leads to normative judgments and the use of pseudoscientific terms like "degradation" that imply a linear, reversible pattern to ecosystem change; alternative nonequilibrium models take into consideration multiple variables and outcomes and focus on processes rather than endpoints. States and transition models may include multiple possible stable states, and the manager must chose among them as a management, or sustainability, goal. Second, the move to broader use of state and transition models is limited by the scarcity of available information and the need for intensive research. While organizations like the US Department of the Interior's Natural Resources Conservation Service are linking state and transition models to soil surveys and ecological sites throughout the United States, there is still a paucity of data from which to build and rigorously test these models. That said, the few tests of state and transition models have shown the predictive value of the models when supported by site- and time-specific data. Third, adequate monitoring of ecosystem response to environment and management may be too costly without the development of new and efficient methodologies.

Lynn HUNTSINGER and James W. BARTOLOME
University of California, Berkeley

See also in the *Berkshire Encyclopedia of Sustainability* Biodiversity; Biogeography; Community Ecology; Complexity Theory; Dam Removal; Disturbance; Ecological Forecasting; Fire Management; Food Webs; Forest Management; Plant-Animal Interactions; Population Dynamics; Regime Shifts; Resilience; Rewilding

FURTHER READING

Allen-Diaz, Barbara H., & Bartolome, James W. (1998). Sagebrush-grass vegetation dynamics: Comparing classical and state-transition models. *Ecological Applications, 8*(3), 795–804.

Bartolome, James W.; Jackson, Randall D.; & Allen-Diaz Barbara H. (2009). Developing data-driven descriptive models for California grassland. In R. J. Hobbs & K. N. Suding (Eds.), *New models for ecosystem dynamics and restoration* (pp. 124–135). Washington, DC: Island Press.

Begon, Michael; Harper, John L.; & Townsend, Colin R. (1996). *Ecology: Individuals, populations, and communities* (3rd ed.). Cambridge, MA: Blackwell Science.

Clements, Frederic E. (1916). Plant succession: An analysis of the development of vegetation. In *Carnegie Institution of Washington Publication No. 520*. Washington, DC: Carnegie Institution of Washington.

Connell, Joseph H., & Slatyer, Ralph O. (1977). Mechanisms of succession in natural communities and their role in community stability and organization. *The American Naturalist, 111*, 1119–1144.

Cramer, Viki A., & Hobbs, Richard J. (Eds.). (2007). *Old fields: Dynamics and restoration of abandoned farmland*. Washington, DC: Island Press.

Egler, Frank E. (1954). Vegetation science concepts. I: Initial floristic composition—A factor in old-field vegetation development. *Vegetatio, 4*, 412–417.

Fairhead, James, & Leach, Melissa. (1995). False forest history, complicit social analysis: Rethinking some west African environmental narratives. *World Development, 23*(6), 1023–1035.

Huntsinger, Lynn, & Bartolome, James W. (1992). A state-transition model of the ecological dynamics of *Quercus* dominated woodlands in California and southern Spain. *Vegetatio, 99–100*, 299–305.

Odum, Eugene P. (1963). *Ecology*. New York: Holt, Rinehart & Winston.

Pignatti, Sandro, & Ubrizsy Savoia, Andrea. (1989). Early use of the succession concept by G. M. Lancisi in 1714. *Vegetatio, 84*, 113–115.

Tillman, David. (1985). The resource-ratio hypothesis of plant succession. *The American Naturalist, 125*(6), 827–852.

Westoby, Mark; Walker, Brian; & Noy-Meir, Imanuel. (1989). Opportunistic management for rangelands not at equilibrium. *Journal of Range Management, 42*(4), 266–274.

Wiens, John A. (1984). On understanding a non-equilibrium world: Myth and reality in community patterns and processes. In D. R. Strong Jr., D. Simberloff, L. G. Abele & A. B. Thistle. (Eds.). *Ecological communities: Conceptual issues and the evidence* (pp. 439–457). Princeton, NJ: Princeton University Press.

Yates, Colin J., & Richard J. Hobbs. (1997). Woodland restoration in the Western Australian wheatbelt: A conceptual framework using a state and transition model. *Restoration Ecology, 5*(1), 28–35.

True Cost Economics

Traditional economics doesn't account for prevailing business practices' collateral damage to the health and well-being of humankind and ecosystems; critics think the free market / cost-price system is detached from these realities and undermines sustainability. True cost economics attempts to incorporate environmental and health damage into product pricing, which could influence consumption patterns and allow the burden to be borne more equitably.

As prepared by the anti-consumerist, not-for-profit organization AdBusters, the "True Cost Economics Manifesto" begins: "We, the Undersigned, make this accusation: that you, the teachers of neoclassical economics and the students that you graduate, have perpetuated a gigantic fraud upon the world" (Bauwens 2009). The alleged fraud consists of conjuring the illusion of perpetual progress and endless growth from theoretical abstractions that obscure a real world of accelerating ecological decay and widespread human misery. One possible wake-up call is the reality check implicit in true cost economics.

The theoretical debate around true cost economics has been simmering for decades but has now begun to boil over. Its proponents believe that the neoclassical (or neoliberal) economics that has dominated the world for at least the past fifty years is hopelessly flawed. Neoclassical free-market models are detached from physical or social realities. They float disdainfully above the ecological, cultural, and ethical contexts within which the real economy is embedded and thus undermine the quest for sustainability.

Neoclassical economists have traditionally been content to allow the prices of goods and services to be determined solely by the law of supply and demand. However, in unregulated markets, only direct producer costs (for rent, labor, resources, and capital, for example) are reflected in consumer prices. The prevailing cost-price system does not account for the collateral damage to ecosystems, human communities, or population health caused by many production processes. These external (outside the market) costs are born disproportionately by third parties or society at large—and, of course, the ecosphere. Because negative externalities represent real costs, the goods and services inflicting them enter the marketplace at prices below their true cost of production. Such underpricing leads to overconsumption, inefficient resource use, and pollution—all classic symptoms of market failure.

True Cost Pricing

By contrast, in a true cost economic system, consumer prices would incorporate environmental, health, and other welfare damage costs of production. When prices "tell the truth" about costs, consumers adjust their consumption patterns accordingly, purchasing fewer ecologically costly goods. Markets would operate more efficiently, producers would innovate and adopt cleaner production processes, total production/consumption would decline (a good thing in a resource-stressed world), pollution and health costs would be would be reduced to insignificance, and third parties would be relieved of an unfair burden.

With so much going for it, why hasn't true cost economics become standard economics? As we shall see, the answer is complicated but, for starters, consider that the true cost approach would result in steep increases in prices for many goods and services that are today within reach of even lower-income groups. For example, some analysts suggest that true cost economics would raise the price of an average car by many thousands of dollars at current production practices. Who would vote for that? Ours is a global consumer culture accustomed to getting more and more for less and less. Correcting for market failure requires government intervention, and any policy that spawned dramatic price increases would spell electoral disaster for the governing party.

Incentive-Based Instruments

Economists have long argued about how best to internalize wayward external costs. By the 1960s, two main schools of thought had coalesced around the competing theories of Arthur Cecil Pigou and Ronald Harry Coase.

Pigovian Taxes

The English economist Arthur Cecil Pigou (1932) argued in *The Economics of Welfare* (first published in 1920) that the existence of externalities justifies government action. He advocated that pollution charges or taxes be applied to offending activities, to better reflect their true social costs and reduce consumption of the relevant goods. (Pigou also suggested that government subsidize private activities generating positive externalities. This would encourage private engagement in those activities and enhance the gains to society at large.)

The mechanics are simple. Imagine an economy in which a number of polluting industries are imposing unaccounted "pollution avoidance costs" on other industries (such as extra expenses for air and water treatment) as well as various "welfare damage costs" on the public (health costs, aesthetic losses, and forgone recreational opportunities, for example). A reasonable public policy objective would be to "internalize these externalities," keeping in mind that, in a total social-cost framework, any solution that imposes pollution-prevention costs on the polluters greater than the anticipated benefits (avoided costs) to other firms and the public would be inefficient.

Government could, in theory, meet this objective through a flat pollution tax per unit of contaminant emitted. The tax would force each polluting firm to decide between treating its wastes and paying the tax. Acting rationally, firms would opt to treat their emissions to the point where their rising, marginal unit-treatment costs just equal the tax. Beyond that, it would be cheaper to pay the tax.

Since different firms have differing "marginal cost of treatment" curves, each will treat a different proportion of its wastes. However, since the tax is uniform, the marginal cost at which firms switch to paying the tax will be the same for all. This ensures that low-cost polluters do most of the cleanup and minimizes the total costs of treatment (a necessary condition for maximum efficiency).

Note that pollution taxes bring the maximum amount of private information to bear on cleanup strategies.

Government need not know the internal processes or cost structures of affected firms. Taxes also pay for their own administration and enforcement. But there is a major problem: in the absence of perfect knowledge, setting the tax rate is just an educated guess. If set too low, the tax will not induce sufficient waste treatment; if too high, firms will inefficiently overtreat their emissions (that is, marginal treatment costs will exceed welfare gains). Subsequent corrective adjustments to the tax are both materially—and politically—costly.

Coasian Bargaining

Despite such drawbacks, the logic of Pigovian taxes charmed most economists until 1960, when it was seriously challenged by the economist Ronald Coase, who argued that if property rights to resources (including sink capacity, or nature's ability to absorb humanity's output and wastes) were clearly defined, then government intervention to correct for externalities was unnecessary, and society could avoid the administrative and enforcement costs of antipollution taxes.

In the absence of significant transaction costs, both polluters and affected parties have a financial incentive to reach an efficient solution through bargaining, regardless of who owns the contested resource. Suppose you, as a papermaker, hold the right to pollute a stream from which I, as a food processor, draw my water. Then I have an incentive to pay you to treat your waste water as long as the cost to me is less than the cost of treating my intake water. Similarly you have an incentive to accept payment because you can profit from treating your wastes. This is because the marginal cost to you of decontaminating your relatively concentrated wastewater is less than the marginal cost to me of cleaning up my more diluted intake water. Of course with more intensive treatment your marginal costs rise to the point where I would save money by shifting to treating my intake water. We bargain around this point—you to maximize receipts, me to minimize pollution avoidance costs.

Now assume I hold the rights to the stream. You have an incentive to pay me to let you pollute as long as my unit price is less than your marginal wastewater treatment costs; I would profit by selling you pollution rights, but only to the point where your payments just cover the cost of treating my now-contaminated intake water. Again, we negotiate an agreement that internalizes costs with no government involvement.

But what if there are thousands of competing firms and myriad other social entities with interests in the negotiated outcome? In the real world, the initial allocation of property rights does matter, and total transaction costs (for research and information, negotiation, administration, and so on) balloon astronomically. Moreover, it is naïve to think that any collection of narrowly "self-interested utility maximizers" will arrive at a solution that is optimal for nature or society at large. As ecological economist Herman Daly constantly reminds us, the "self" in which we are primarily interested is not an isolated atom, but is defined by its relationships in community and by diverse biophysical connections that are affected, but not acknowledged, in economic transactions. This negates any possibility of achieving an efficient or effective solution through private bargaining alone—community (or government) must be involved.

Cap-and-Trade Systems

Government does have one attractive policy option that capitalizes on financial incentives and the allocative efficiency of competitive markets to take the guesswork out of pollution pricing. So-called cap-and-trade schemes also combine fixed emission levels with tradable pollution rights and thus separate the public-policy issue of what constitutes acceptable environmental quality from the legitimate economic question of efficient allocation. The US Clean Air Act was based, in part, on caps and tradable permits, as is the early Obama administration's proposed climate change policy (the Waxman-Markey bill).

In an ideal cap-and-trade scenario, government solicits scientific advice and public opinion to set desirable environmental quality objectives and place a firm limit (cap) on allowable emissions that reflects regional assimilation capacity. The allowable emissions are then divided into a fixed number of shares or permits and distributed by some fair means to existing polluters. After initial distribution, subsequent allocation is determined by trading in an open market. The price per share is thus set by the usual law of supply and demand, except that supply is fixed. If demand increases, prices rise, inducing market participants to invest in more efficient (cheaper) production or waste-treatment processes and reduce their need for pollution rights. New businesses, or firms needing additional shares, purchase them from businesses that no longer require their full quota. Inefficient players are forced out of the market.

In theory then, a tradable permit scheme could achieve ecological and social objectives with certainty, using multiple sources of public and private information to set limits and prices while internalizing erstwhile externalities with market efficiency. And if government charged for the initial distribution of permits or demanded a royalty on subsequent trades (environmental capacity is, after all, a public good), the system would pay for its own implementation and monitoring.

Invisible and Intangible Costs

We have already shown that, despite their theoretical appeal, standard Pigovian and Coasian approaches have serious practical weaknesses. We now consider a fundamental problem that plagues all formal instruments for true cost economics—identifying and monetizing intangible and invisible costs.

Direct production costs and external property damage costs are readily determined from current market prices. But there are no markets for numerous indirect use, non-use, option, and existence values associated with ecosystems and communities. The market price for a truckload of logs, for example, is mute about the flood control, water purification, biodiversity, carbon sink, and aesthetic and spiritual values sacrificed in clear-cutting the forest. This is why the consumer purchasing a board foot of lumber—or just about anything else—doesn't come close to paying the full social cost of production.

One problem is that assigning a valid money price to something assumes the ability to compress all the values associated with that thing into a single metric. Arild Vatn and Daniel Bromley (1994) identify three theoretical obstacles to such inclusive pricing of environmental (or social) entities:

- A *cognition problem* always exists in the absence of perfect knowledge, and the simple fact is that many critical functions of species and ecosystems are cognitively invisible. This "functional transparency" means that the cost of losing any important element of an ecosystem may be unknowable until that element has been destroyed. We obviously cannot place any value on that which we cannot know.
- An *incongruity problem* exists when the values associated with an ecologically significant good are incongruous or incommensurable with dollar values. How can we conflate the market price of duck breast with the sheer aesthetic rush experienced from witnessing a wedge of mallards in full flight over the marsh?
- A *composition problem* arises because in ecosystems the whole may be dependent on each of its fundamental parts. This means that value of any single component (for example, a species or nutrient) cannot be interpreted independently of the value of the whole.

These and related barriers mean that mainstream efforts to derive accurate, unambiguous *money* values for complex ecological entities (such as contingent valuation) are doomed to failure—we cannot compute costs. Our assumed ability to commoditize nature and basic life support is an arrogant fiction (and in any event, may not be such a good idea).

Transcending Benefit-Cost Analysis

All important decisions involve weighing the relative gains and losses associated with the various options under consideration. As we have seen, pollution charges, one-on-one bargaining, and cap-and-trade schemes all force the affected parties to compute private self-interested benefit-cost ratios as the basis for their internal waste management decisions ("Do we treat our wastes or pay the tax?").

More generally, formal benefit-cost analysis (BCA) purports to provide a comprehensive comparison of the discounted future benefits and costs associated with different development options. The efficiency goal is to maximize any positive difference between gains and losses. Because of its conceptual simplicity and theoretical elegance, many economists regard BCA as the definitive tool for both private and public policy decision making. In an ideal world BCA would therefore be critical to true cost economics.

But this is not an ideal world—there are practical flies in the theoretical ointment. Missing data and irreducible uncertainty combined with limited resources and ideologically tainted analyses explode any claim that BCA produces a socially optimal true cost outcome. The fact is that comprehensive true cost economics is beyond our analytic reach.

This is no minor glitch. Ignorance of critical ecological and social costs has arguably long biased modern society toward endless growth even as the ecosphere slowly implodes. It is entirely possible that if we could subject the global economy to a valid BCA, we would find that the ecological and social costs of growth at the margin now exceed the benefits. We may have entered an era of what Herman Daly calls "uneconomic growth"—growth that makes the world poorer, not richer. True cost economics may well mean no-growth economics. The fact that the world's rich and powerful reap most of the benefits of growth, while the poor and the global commons bear most of the unaccounted costs, undoubtedly contributes to present policy paralysis.

Such conclusions are not cause for despair but rather should liberate society from the dictates of oppressively wrong-headed economic models. Governments, the private sector, and nongovernmental organizations must learn to eschew "crackpot rigor." We all share this single planet and cannot afford to be blinded by faulty theory and vacuous analysis. By all means, use BCA for those tangible things to which we can legitimately ascribe a dollar value. This may bring us closer to the efficient market economy to which we aspire. But both business and ordinary citizens must recognize that the results are not in themselves a sufficient basis for decision making.

In the end, sustainability is mainly a political, not an analytic, goal. Society must recognize that even as we strive for true cost economics, the most critical ecological and social choices must be made "without prices, without apologies" (Vatn and Bromley 1994). Given the scale of the problem, multiple conflicting values, gross distributive inequity, and a deepening well of uncertainty, there is no substitute for informed, cautiously practical political judgment, all for the common good.

William E. REES

University of British Columbia

See also in the *Berkshire Encyclopedia of Sustainability* Accounting; Cap-and-Trade Legislation; Consumer Behavior; Development, Sustainable; Ecolabeling; Ecosystem Services; Energy Efficiency; Green GDP; Human Rights; Investment, Socially Responsible (SRI); Natural Capitalism; Performance Metrics; Triple Bottom Line

FURTHER READING

Baumol, William J., & Oates, Wallace E. (1988). *The theory of environmental policy* (2nd ed.). New York: Cambridge University Press.

Bauwens, Michel. (2009, August 9). True cost economics manifesto. Retrieved August 26, 2009, from http://blog.p2pfoundation.net/the-true-cost-economics-manifesto/2009/08/09

Coase, Ronald H. (1960). The problem of social cost. *Journal of Law and Economics, 1*(3), 1–44. Retrieved August 31, 2009, from http://www.sfu.ca/~allen/CoaseJLE1960.pdf

Dales, John. (1968). *Pollution, property and prices: An essay in policy-making and economics*. Toronto: University of Toronto Press.

Daly, Herman E. (1981). *Steady-state economics* (2nd ed.). Washington, DC: Island Press.

Daly, Herman E., & Cobb, John B., Jr. (1989). *For the common good: Redirecting the economy toward community, the environment, and a sustainable future*. Boston: Beacon Press.

Daly, Herman E. & Farley, Joshua. (2004). *Ecological economics: Principles and applications*. Washington, DC: Island Press.

Daly, Herman E., & Townsend, Kenneth N. (1993). *Valuing the Earth: Economics, ecology, ethics*. Cambridge, MA: MIT Press

Hahn, Robert W. (1989). *A primer on environmental policy design*. Chur, Switzerland: Harwood Academic Publishers.

Jacobs, Michael. (1991). *The green economy: Environment, sustainable development, and the politics of the future*. London: Pluto Press.

Lave, Lester B., & Gruenspecht, Howard K. (1991). Increasing the efficiency and effectiveness of environmental decisions: Benefit-cost analysis and effluent fees—a critical analysis. *Journal of the Air and Waste Management Association, 41*(5), 680–693.

Manno, Jack P. (2000). *Privileged goods: Commoditization and its impact on environment and society*. Boca Raton, FL: Lewis Publishers.

O'Neill, John. (2006). *Markets, deliberation and environment*. London: Routledge.

O'Neill, John; Holland, Alan; & Light, Andrew. (2008). *Environmental values*. London: Routledge.

Pearce, David W. (1993). *Economic values and the natural world*. Cambridge, MA: MIT Press.

Pigou, Arthur C. (1932). *The economics of welfare* (4th ed.). London: MacMillan,

Prugh, Thomas; Costanza, Robert; Cumberland, J. H.; Daly, Herman E.; Goodland, Robert; & Norgaard, Richard B. (1995). *Natural capital and human economic survival*. Solomons, MD: ISEE Press.

True cost economics. (n.d.). Retrieved November 10, 2009, from http://www.investopedia.com/terms/t/truecosteconomics.asp

Themes, Brendan. (2004, August 26). True cost economics. Retrieved August 26, 2009, from http://www.utne.com/2004-08-01/TrueCostEconomics.aspx

Vatn, Arild, & Bromley, Daniel W. (1994). Choices without prices without apologies. *Journal of Environmental Economics and Management, 26*(2), 129–148.

Victor, Peter A. (2008). *Managing without growth: Slower by design, not disaster*. Cheltenham, UK: Edward Elgar.

Water

In the twenty-first century, societies struggle to manage water resources, whether for agriculture, industry, energy, recreation, or consumption. Sustainable goals include protecting endangered species, preserving wilderness, and allowing natural systems to function without human intervention. Of water's many uses, some complement each other and environmental protection while others do not—both nutrients and pathogens, for instance, are transported by water. Defining and sorting out water rights will be a prime challenge of the future, for citizens and water managers alike.

Humankind's ancient, massive, and growing intervention in natural hydrological processes, including capturing and rerouting flows of water and altering water quality, has resulted in improved human health and longevity, and has created opportunities for humans to live in places, numbers, and comfort that otherwise wouldn't be possible. Between 1700 and 1900, human withdrawals of freshwater from aquifers, rivers, and lakes increased five times, from 110 cubic kilometers to 580 cubic kilometers. Over the next hundred years, withdrawals increased another nine times to 5,190 cubic kilometers, which is more than 10 percent of the Earth's available flows. Although the same water can be used repeatedly, different uses require different levels of quality, quantity, and flow characteristics. In the twenty-first century, societies are grappling with how to manage water resources to achieve multiple goals. Environment-related goals include protection of endangered species, provision of recreational and wilderness experiences, and allowing natural systems to provide services that humans would otherwise have to provide themselves.

Physical and Biological Contexts

The ongoing exchange of seawater and freshwater is called the hydrologic cycle, and is powered by the sun's energy causing evaporation. As evaporated water cools in the upper atmosphere, it condenses and falls as rain or snow. Although most evaporation and precipitation occur over oceans, clouds are also carried inland by winds, bringing the water that fills rivers, recharges groundwater aquifers, and rejuvenates the Earth's terrestrial biomass. The cycle continues as the water eventually evaporates again.

Key physical aspects of water enable it to play its central role. Substances can dissolve or be suspended in water. This enables flowing water to transport them from one place to another. Rivers carry nutrients from mountains to valleys to soils in flood plains. Poisons and pathogens also reside in water, making it a common source of disease to humans and other species.

Evaporating water leaves minerals behind. Unless replenished with a new inflow, the minerals left behind become more concentrated, rendering the remaining water less fit for use. Water can impart energy both in terms of its elevation (falling water drives hydropower turbines), and in terms of heated water molecules (the evaporation process transfers energy from liquid water to the atmosphere). But the energy in water also propels floods, one of Earth's most devastating natural disasters. Since the hydrogen side of a water molecule is more positively charged and the oxygen side more negatively charged, liquid water molecules tend to align their positive and negative poles and pull together. This property enables water to flow from roots to leaves of plants, and enables water to serve as a solvent. Finally, water freezes from the top down, allowing fish and other species to survive the winter in unfrozen depths, which has enabled more long-lived and complex organisms and food webs to emerge.

Water's origins on Earth remain a mystery. It may have been released from within the floating rocky debris that first formed the Earth, or it may have arrived little by little in the form of icy meteors striking the dry young planet. Now twice as much of the Earth is covered by water as is covered by land. Ninety-seven percent of all of Earth's water is found in oceans, and 2 percent is located in polar and mountain ice. Of the remainder, 95 percent is located in underground aquifers, and the rest can be found in lakes, inland seas, surface soils, the atmosphere itself, living biomass, and stream channels. The ironic phrase "as predictable as weather" summarizes the variable nature of water's stocks and flows: wet and dry periods, droughts, and floods all are common occurrences, though their exact timing and magnitude cannot be predicted.

Freshwater ecosystems contain a large proportion of the world's biodiversity compared to their size. Making up just 1 percent of the Earth's surface, they serve as home to 40 percent of the world's fish species, while 12 percent of animal species reside in fresh water. With respect to oceans, the Food and Agriculture Organization of the United Nations (FAO) estimates that only around 15 percent of currently exploited marine species can support an increase in catch rates over the long term (FAO 2011, 8).

Wetlands—regions where the prevalence of water results in saturated soils and that support life adapted to saturated soil conditions—are among the world's most biologically productive. They also provide such beneficial services to humans as filtering water, slowing and capturing flood flows, and serving as fish and wildlife habitat. Biological processes in wetlands are mimicked in today's wastewater treatment facilities, which concentrate and accelerate the natural processes that break down and remove solids and pathogens from fresh water.

Irrigation

Irrigation remains the leading human use of water, accounting for 70 percent of all freshwater consumption. Forty percent of all crops produced today have been irrigated (Sundquist 2007). Irrigated agriculture may have originated with the Sumerians of Mesopotamia around 4000 BCE. By 3200 BCE, Egyptians were digging irrigation canals off of the Nile River. As of 700 BCE, *qanats*—deep tunnels that deliver water from an underground aquifer to a farming region—were well established in Persia. The Sabeans living in Marib (modern-day Yemen) built a large agricultural diversion dam in roughly 500 BCE. In North America, by 200 BCE the Hohokam tribes located in current-day Arizona had established extensive irrigation canals drawing from the Gila and Salt rivers. By 1900, roughly 40 million hectares were irrigated worldwide. This quantity grew seven-fold in the twentieth century to 270 million hectares. Irrigation particularly accelerated after 1950 with the introduction of "Green Revolution" high-yielding seed varieties that required extensive and well-regulated inputs of water.

The environmental impacts of irrigation canals include reducing the amount of water available for instream ecological processes and for riparian (at or near the river bank) wetlands, and altering riparian zones in the places where diversion works exist. Further environmental impacts occur along canal routes where dry terrestrial ecosystems are transformed into flowing water systems, terrestrial habitats are fragmented, and migratory paths of terrestrial animals are blocked. Although irrigation canals do support wildlife, the flow regime of canals is keyed toward agricultural production, not a region's natural hydrological processes, creating adaptation challenges for local water-dependent species. While irrigated agriculture may increase overall biotic productivity, especially in dry regions, it may also reduce a region's biodiversity as farmers devote fields to a few or a single food-crop species (called monocropping).

In regions where salts naturally occur in the soil column, intensive irrigation can lead to soil salinization. As irrigation water rises to the surface and evaporates or transpires, remaining salts concentrate, ultimately rendering the root zone infertile. Soil salinization is blamed for the destruction of numerous early agriculture-dependent civilizations of Mesopotamia and possibly South America. Today, roughly 20 percent of irrigated acreage worldwide has elevated quantities of salts (Munns 2004).

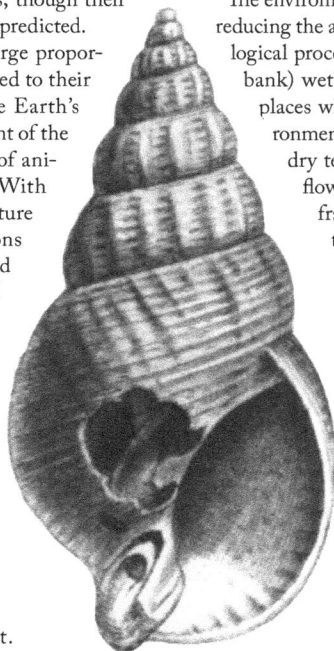

Hydropower

Waterwheels are among humankind's earliest machines, being utilized by the early Greek, Egyptian, and Chinese civilizations. The earliest waterwheels were used to grind grain and to lift water out of a river into another channel. Over the centuries, other applications emerged, with over 500,000 water wheels existing in Europe by 1800. Hydropower consisted of transforming the movement of water directly into the movement of machinery until 1882 when the world's first hydroelectric plant was built on Wisconsin's Fox River. The growing demand for electricity spurred by the industrial era helped launch an era of large dam building worldwide. As of 1900, roughly 700 dams higher than 15 meters had been commissioned, with 220 located in England. China was the leading builder: at the time of its 1949 revolution, it had 23 large and medium-scale dams; by 1990, over 80,000 had been built (Fu 2007). Although the exact number of dams in the world is not known, according to the World Commission on Dams (WCD) there are about 45,000 dams higher than 15 meters in height worldwide (WCD 2000). The number of dams worldwide lower than 15 meters is not well known, but estimates are in the millions (Smith 1971, WCD 2000).

The largest hydropower dams, such as the Three Gorges Dam on the Yangzi (Chang) River, create reservoirs that dramatically alter the ecology of river systems as well as displace large numbers of people. Above these large dams, flowing water regimes become standing water regimes, and riparian habitat is flooded. Shorelines of reservoirs often are barren since plant and animal life are not adapted to ever-shifting reservoir elevations determined by hydropower, agricultural, urban, downstream environmental, and recreational needs. Dams block sediment flows, causing siltation above the dam, alteration or loss of habitat for bottom-dwelling (benthic) species, and a reduction in the deposition of nutrients in the floodplain. Dams also block passageways for migratory fish. By retaining flood flows in reservoirs, dams de-link the main channels of rivers from nearby wetlands, reducing the contribution each makes to ecological processes occurring in the other. Water temperature increases behind dams due to prolonged exposure of standing water to sunlight, which can degrade downstream fish spawning habitat when the warmer water is released.

There have been profound indirect environmental effects of hydropower and large dams. Hydropower provided a large proportion of the energy that fueled the Industrial Revolution. In the early 1900s, 40 percent of all US electricity was generated by hydropower. By 2010, hydropower provided roughly 7 percent of US electricity and over 16 percent of the world's electricity (Egan 2011). Large dam/reservoir systems enabled vast regions to transition from undeveloped wetland ecosystems to agricultural ecosystems. By 1980 California had lost 91 percent of its original wetlands largely as a result of the transformation of its great central valley from wetlands to agriculture. This substantially reduced available habitat and population sizes of numerous bird, terrestrial, and aquatic species.

Transportation

Rivers have served as commercial highways since ancient times. Civilizations have built dams and canals both to stabilize the elevation of waterways and to extend their reach. Rivers, including the Mississippi and the Rhine, have been dredged and straightened to permit the passage of longer barges with deeper drafts. Obstructions such as submerged boulders, branches, and tree trunks have also been removed from river channels. Environmental impacts include disturbing biotic activity in the benthic (river bed) zone due to dredging; curtailing annual flood events; accelerating the velocity and intensity of floods; and reducing shade and protection provided by submerged and partially submerged objects. Over the twentieth century, the number of rivers altered to improve transportation increased dramatically.

Canals also have played crucial roles in extending the range of human habitation and increasing the intensity of economic activity on frontiers. Canals enable building materials and machinery to be transported to frontiers while also enabling crops and natural resources to be shipped cheaply and quickly back to urban centers. China's 2,500-year-old Grand Canal extends over 1,600 kilometers enabling extensive agriculture and other industry to occur far from natural rivers. Completed in 1825, North America's 580-kilometer Erie Canal helped link New York Harbor to the Great Lakes, facilitating the rapid westward expansion and industrial growth of the United States.

Flood Control

Floods arise naturally due to heavy rainfall and/or rapid snowmelt, and on rare occasions result from catastrophic dam failure. Some 10,000 floods occur each year worldwide. Modern history's worst recorded flood occurred in 1887 along China's Huang (Yellow) River. A million people died as the waters engulfed eleven large towns and hundreds of villages. In 2010, flooding on the Indus in the Pakistani provinces of Khyber Pakhtunkhwa, Punjab, and Sindh caused more than 1,500 deaths and left an estimated 6 million people homeless.

Environmental impacts of flooding include transfer of sediments and nutrients from uplands to the floodplain; creation of temporary aquatic links between rivers and nearby lakes and wetlands enabling the relocation of aquatic species; and re-routing of rivers into alternate or new channels. Floods help make soils fertile and aid forest growth. Levees (embankments that confine a flood), while protecting the immediate area from flooding, often serve to accelerate floodwaters, increasing their potential for devastation downstream. Levees also alter flood-dependent ecological processes where they are located. Restoration of upstream wetlands and forests, which serves to slow down floodwaters, became a part of flood-management planning in the 1990s.

Water for Cities

The ancient Romans took immense pride in their aboveground aqueducts, which by 300 BCE were delivering more than 1.5 million liters of water per day to Roman citizens. Roman cities also included extensive systems of pipes that delivered water from the aqueducts to houses and baths, as well as separate systems that carried sewage wastes away from homes. Today, roughly 87 percent of all urban dwellers have access to a fresh water supply, and 62 percent have access to sanitation facilities (WHO and UNICEF 2010).

Throughout human history, societies have utilized rivers to carry wastes away, increasing the risk of contamination downstream. Water quality impacts can be divided between biological contaminants and chemical contaminants, the importance of the latter growing since the dawn of the industrial era. Though some ancient societies, such as the Chinese, adopted practices that separated human wastes from waters intended for human consumption, until the mid-nineteenth century, links between history's many plagues and the human relationship with water were only vaguely understood. Water-related diseases such as typhoid, cholera, and malaria took their toll worldwide. As recently as 1885, a cholera epidemic killed over 80,000 people in Chicago. Advances in epidemiology (the study of diseases in large populations) also occurred during this period. In 1854, Dr. John Snow linked London's cholera outbreaks to a public well that drew water from the Thames River just downstream from the city's raw sewage outfalls. His discovery launched investigations in England and elsewhere into alternative designs for wastewater treatment facilities.

In the twentieth century, as growing cities in Europe and eastern North America increasingly released concentrated biological and industrial wastes into waterways, severe environmental degradation resulted. Biological impacts included a reduction in the quantity of dissolved oxygen in the rivers as living microbial matter multiplied and consumed biological wastes. Oxygen depletion led to die-offs of fish populations.

Pollution generated by industrial and agricultural activities often cannot be broken down by natural processes. Trace elements (including heavy metals), pesticides, petroleum, and petroleum byproducts may leach from mining or farming operations or be flushed away from metal-plating, chemical, refining, and other facilities. They can be damaging or fatal to fish and other aquatic species. Some elements, including mercury and arsenic, bio-accumulate along food chains, reaching concentrations dangerous to humans and birds who eat fish caught in polluted waters.

Water for the Environment

Since the 1970s, industrialized nations have taken significant steps to reduce the loading of industrial pollution into rivers and lakes, including in the United States the passage of the Clean Water Act of 1972. The environmental health of European and North American waterways has improved, symbolized by the capture of an Atlantic salmon in a Rhine River tributary in 1990, twenty-two years after the previous catch occurred. Developing-world waterways remain under environmental pressure from untreated industrial and urban effluents, both of which continue to grow in quantity.

Internationally, *The Convention on Wetlands of International Importance Especially as Waterfowl Habitat*, or Ramsar Convention, of 1971, maintains a list of "Wetlands of International Importance" that by 2011 included 1,951 sites covering 190 million hectares (Ramsar Convention on Wetlands 2011). A growing number of national park systems worldwide also provide protection to aquatic ecosystems located within their boundaries.

Of the many uses of water, some are complementary to each other and to environmental protection while others are not. Sorting out water rights—what they are, who has them, and how they are transferred—will be a prime challenge for years to come for citizens and water managers alike.

Brent M. HADDAD
University of California, Santa Cruz

See also in the *Berkshire Encyclopedia of Sustainability* Agriculture (*several articles*); Aquifers; Dams and Reservoirs; Desalination; Fish; Glaciers; Parks and Preserves—Marine; Oceans and Seas; Rivers; Water Energy; Wetlands

FURTHER READING

Cech, Thomas V. (2003). *Principles of water resources: History, development, management, and policy.* New York: John Wiley & Sons.

Dahl, Thomas E. (1990). Wetlands losses in the United States: 1780s to 1980s. Washington, DC: US Department of the Interior, Fish and Wildlife Service. Retrieved September 12, 2011, from http://www.npwrc.usgs.gov/resource/wetlands/wetloss/

Egan, John. (2011). Clean and green hydropower needs to maintain its share in US electricity mix. Retrieved September 19, 2011, from http://www.marketwatch.com/story/clean-and-green-hydropower-needs-to-maintain-its-share-in-us-electricity-mix-a-navigating-the-currents-of-change-webcast-on-industrialinfo-com-2011-07-13

Food and Agriculture Organization of the United Nations (FAO). (2011). *The state of world fisheries and aquaculture 2010.* Rome: FAO.

Fu, Shui. (2007). A profile of dams in China. Retrieved September 19, 2011, from http://www.internationalrivers.org/china/three-gorges-dam/profile-dams-china

McCully, Patrick. (1996). *Silenced rivers: The ecology and politics of large dams.* London: Zed Books.

McNeill, John R. (2000). Something new under the sun: An environmental history of the 20th century. New York: W.W. Norton.

Munns, Rana. (2004). The impact of salinity stress. Retrieved September 19, 2011, from http://www.plantstress.com/Articles/salinity_i/salinity_i.htm

Ramsar Convention on Wetlands. (2011). Ramsar list of wetlands of international importance. Retrieved September 12, 2011, from http://www.ramsar.org/cda/en/ramsar-documents-list/main/ramsar/1-31-218_4000_0

Smith, Norman. (1971). *A history of dams.* London: Peter Davies.

Sundquist, Bruce. (2007). Chapter 1: Irrigation Overview. In *The earth's carrying capacity: Some related reviews and analysis.* Retrieved September 16, 2011, from http://home.windstream.net/bsundquist1/ir1.html

United Nations Environment Program (2000). Report of the International Task Force for Assessing the Baia Mare Accident.

Wolf, Aaron T.; Natharius, Jeffrey A.; Danielson, Jeffrey J., Ward, Brian S.; & Pender, Jan K. (1999). International river basins of the world. *International Journal of Water Resources Development 15*(4), 387–427.

World Commission on Dams (WCD). (2000). *Dams and development: A new framework for decisions-making.* London: Earthscan.

World Health Organization and United Nations Children's Fund (WHO and UNICEF). (2010). Progress on sanitation and drinking-water: Update 2010. Retrieved September 19, 2011, from http://www.unicef.org/media/files/JMP-2010Final.pdf

Wetlands

Wetlands provide important ecosystem services such as habitat, food, opportunities for recreation, and water supply. The definition of wetlands remains controversial, but the definitions adopted by governments in legislation and policy are critical in determining the scope of management, conservation, or rehabilitation.

For decades, wetlands have been the focus of legislators, land use planners and managers, and nongovernmental environmental and conservation organizations. This reflects their increased understanding of the ecosystem services that wetlands provide (e.g., natural resources, flood regulation, and recreation), which are essential for sustainability at the landscape scale. (The landscape scale is the context in which a wetland is situated and in which interactions between different components of the landscape occur.) This focus also recognizes that, both deliberately and accidentally, human society has been destroying and degrading wetlands for hundreds, and in some cases, thousands of years.

What Are Wetlands?

Wetlands as a term encompasses a wide range of environmental conditions. The limits of this range remain controversial, and different jurisdictions have adopted different definitions. The broadest definition is probably that taken in the Convention on Wetlands of International Importance especially as Waterfowl Habitat—one of the oldest international environmental treaties, signed in 1971 and generally referred to as the Ramsar Convention after the Iranian city where the document was conceived. Article 1.1 of the Convention defines wetlands as "areas of marsh, fen, peatland or water, whether natural or artificial, permanent or temporary, with water that is static or flowing, fresh, brackish or salt, including areas of marine water the depth of which at low tide does not exceed six metres" (Ramsar Convention 1971).

The convention imposes a number of obligations on the contracting parties, including that each country nominates sites to be included on the Ramsar List of Wetlands of International Importance and that each adopts a principle of wise use for all wetlands within its jurisdiction. (*Wise use* is interpreted here as requiring wetlands to be managed sustainably; this is different than the anti-environmentalist movement in the US West by the same name). For purposes of listing important wetlands, countries "may incorporate riparian and coastal zones adjacent to the wetlands, and islands or bodies of marine water deeper than six metres at low tide lying within the wetlands" (Ramsar Convention 1971).

While the convention's definition has terms that refer to specific types of wetlands (marsh and fen), its interpretation is much wider than when these words are used in the vernacular. Importantly, the Ramsar Convention explicitly applies to artificial wetlands (and, indeed, a number of artificial wetlands appear on the Ramsar list primarily because of their value as bird habitat) and to temporary wetlands. Some temporary wetlands are regularly, predictably wet, such as the vernal (spring) pools in California; others may be wet irregularly, such as wetlands in Australia that may only be wet once or twice a century. The latter group includes some of the most extensive wetlands in the world, the normally dry lakes in the arid and semiarid zones. Wetting of these lakes promotes a great flush of biological activity, including supporting, for brief periods, vast numbers of waterbirds. The public rarely thinks of these arid-zone wetlands as wetlands, and yet they are essential for sustaining major components of biodiversity. Even less likely to be thought of as wetlands, and not often captured by wetland policy and legislation, are the waters under the Earth. Subsequent conferences by the parties to the Ramsar Convention have made it clear that "karst and other subterranean hydrological systems" are included within the wetland definition (Ramsar Convention 1971, Resolution VI.5).

Why is groundwater important to a discussion on sustainability? First, groundwater resources provide the daily water supply for much of the world's human population, and as the population grows, groundwater will become even more important. Groundwater is an internationally traded commodity, as bottled "natural" mineral water has become a perceived essential to life in Western society. Nevertheless, we have damaged many aquifers through pollution, and if the human population continues to expand without provision for adequate wastewater treatment and the control of industrial discharges, then the threats to groundwater will increase.

Second, we have become aware that groundwater provides habitats for a great diversity of organisms. Bacterial communities have been found at great depths in the minute cracks and voids in rocks, and a diverse fauna inhabits cave systems and the interstices between rock particles. Although we knew of the existence of this stygofauna from the nineteenth century, it is only within the last few decades that we have discovered that this diverse and highly specialized subterranean fauna occurs globally (Humphreys 2006). Stygofauna's typical characteristics, or specializations, include being colorless and lacking eyes. This fauna is at risk from pollution and water abstraction (i.e., the temporary or permanent removal of water). Not only is groundwater a habitat, but it also provides the water supply for many ecological communities on the Earth's surface, for example, springs and some forests and shrublands. We are only beginning to appreciate the extent of these groundwater-dependent ecosystems and their sensitivity to changes in their groundwater supply.

In many places, people still consider groundwater to have an endless supply, and they therefore have little consideration for its sustainability. Thus rates of extraction may exceed rates of recharge, so the resource is being depleted. In some cases, ancient water is being extracted so that the resource is essentially being mined, with restoration not possible within human time frames. (For example, water being extracted in the twenty-first century for mining operations in Central Australia fell as rain thousands of years ago and then commenced its long, slow journey through the aquifer.)

Finally, in particular locations, groundwater is essential for tourist attractions that are the mainstay of local economies. Sustainability is important to maintaining attractions such as massive cave systems or hot springs and geysers.

Importance of Wetlands

Human societies have long depended directly on wetland resources for their survival. As we developed technologies to drain and modify wetlands on an increasingly large scale, local communities were displaced or forced to adapt to changed circumstances. Wetlands were increasingly viewed as wastelands, and wetland destruction came to be regarded as a public benefit, frequently referred to as *reclamation*, with its connotations of restoring the wetlands to some better state.

The knowledge of those who previously depended on wetlands was lost or ignored, and it is only since the mid-twentieth century that we have a renewed recognition of wetlands' values. The preamble to the Ramsar Convention recognizes these values as the "fundamental ecological functions of wetlands as regulators of water regimes and as habitats supporting a characteristic flora and fauna." (The Oxford English Dictionary [2010] defines regime as "the set of physical conditions and influences to which a system is subject or by which it is maintained"—hence a water regime would include the pattern of rainfall, in terms of amount and temporal distribution, and the relationship between amounts of incoming and outgoing water in a system.)

The definition of wetlands is so broad that it creates difficulties in framing legislation to protect important ecological functions and habitat values. A tropical intertidal mangrove forest and an alpine spring have little in common other than that they both could be characterized as wetlands. Much legislation that deals with wetlands, although attempting to be of general application, actually addresses less than the full range of wetland types that are recognized in the Ramsar Convention, even though the convention's many signatory countries have agreed to promote the wise use of all wetlands.

Wetlands perform many important hydrological functions that include regulating the flow of water through catchments (watersheds); for example, the wetlands in

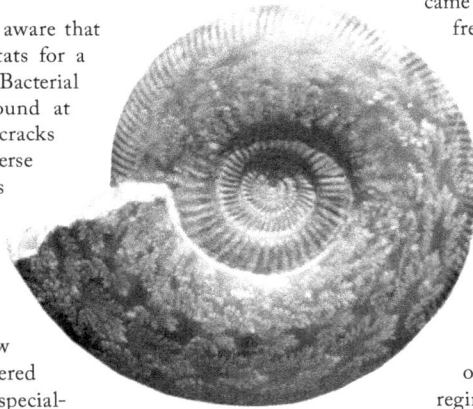

the upper reaches of catchments act as sponges, retaining water that is then released relatively evenly over time. Loss of these wetlands can result in much more variable flows downstream. Floodplain wetlands provide flood storage capacity during times of high flow. Filling and development of these wetlands shifts floodwater elsewhere, including to developed areas. Human settlement historically has been concentrated on floodplains, which provide fertile agricultural areas and land that is fairly easy to develop, and they are often part of trading and communication networks. Loss of flood storage capacity, previously provided by wetlands in floodplains, coupled with an expanded human population, means that greater numbers of lives and properties are at risk from flooding. Along the coasts, mangroves and salt marshes absorb wave energy, protecting the hinterlands from erosion and storm damage. Loss of or damage to intertidal wetlands exposes coastal communities to greater risk from natural hazards. The protection these wetlands provide is not absolute, but it is significant. In 2005, the loss of wetlands led to greater damage in the Mississippi delta from Hurricane Katrina than might otherwise have happened, and the loss of mangroves contributed to the damage from the 2004 Asian tsunami. Wetlands have the advantage over engineered structures of providing a range of habitats and, when they are damaged, of being potentially self-repairing.

The role of wetlands extends beyond being sponges, to being filters capable of absorbing and/or assimilating a range of pollutants. Accordingly, one of the selling points for wetland protection policies is that wetlands are "the kidneys of the landscape." While wetlands continue to protect water quality in waterways, this valuable function has been achieved at a cost. An important attribute of many wetlands is the presence of large bacterial populations, including denitrifying species, in the root zone of plants. While nitrogenous inputs in the form of fertilizer runoff and effluent can be returned to the atmosphere, other pollutants are absorbed and retained within the wetlands, with potentially adverse consequences.

The Ramsar Convention stresses the important habitat value of wetlands. The wetland environment is often inimical to the majority of terrestrial species, but wetland species have evolved various adaptations enabling them to thrive. The wetland biota thus consists predominantly of species restricted to wetland habitats and

constitutes a distinctive component of biodiversity. Conservation of biodiversity is a critical requirement for sustainability and further justifies giving high priority to wetland conservation, particularly to their value as habitats for fish and birds.

When the link between the productivity of estuarine and inshore fisheries and coastal wetlands began to emerge in the late 1950s, it was a major contributor to a shift in public opinion about the value of wetlands and the start of legislative protection. Estuarine and inshore fisheries are, globally, a major component of commercial fisheries, providing employment and food for many people.

In the West, these fisheries are also major recreational resources, valued by many voters and political activists. The recreational fishing lobby has been a major influence on the preparation of wetland protection policies, and it actively opposes developments that threaten specific sites. Modern research has added more support to the importance of wetlands for sustaining fisheries. Although some of the early models of the linkages are now seen as overly simplistic, the importance of wetlands for fisheries is widely accepted, not just by fishers. While the modern views were developed from studies in marine environments, indigenous peoples around the world have long known the importance of wetlands for freshwater fisheries.

The two groups of birds most commonly associated with wetlands are waterfowl (ducks, geese, and swans) and shorebirds (waders), which at times may be present in vast numbers; both groups contain migratory or nomadic species. Historically, both waterfowl and some species of shorebirds were hunted as important food items; this still occurs in parts of the world. There is also a considerable interest in the recreational shooting of waterfowl. Recreational shooters have played an important role in protecting wetlands used by waterfowl, especially migrating waterfowl. In some places, recreational shooters are a large and active group, responsible for continuing conservation initiatives and generating considerable economic benefit, but elsewhere shooting is less favored, replaced by ornithology research and ecotourism.

Since the late twentieth century, there has been a growing appreciation of the importance of wetlands in global geochemical cycles, which move chemical

elements around the Earth. In particular, peatlands are a major store of carbon, and the drainage of peatlands leading to oxidation of the substrate could add substantially to the amount of atmospheric carbon dioxide. The most extensive peatlands are in the high northern latitudes; less well known are the deep deposits under tropical rain forests, which are threatened by conversion to agriculture (e.g., rice cultivation or palm oil plantations) and fire.

Many of the values of wetlands relate to ecological processes occurring at landscape scales. While conservation of individual wetland sites is essential, these sites must be seen in the context of whole catchments and even larger scales. Conservation of migratory birds, for example, demands protection of their habitats throughout the migratory pathway. Migratory wetland birds may travel thousands of kilometers between seasons, and some shorebirds travel between hemispheres. Recent technological advances allow satellite telemetry tracking of individual birds, leading to a better understanding of migration. The record flight to date is by a bar-tailed godwit, code-named E7, which was tracked from northern New Zealand, where it spent the Southern Hemisphere's summer, through a staging post in northern Asia in the Yellow Sea, to its breeding grounds in Alaska. Remarkably, at the end of the Northern Hemisphere's summer, it flew 11,680 kilometers nonstop back to New Zealand (Woodley 2009). Sustaining godwits and other shorebirds will require international collaboration. Given the economic value of current and proposed development in the world's coastal zone, this will be a major task.

Loss and Degradation of Wetlands

Human actions have caused substantial loss and degradation of wetlands, affecting their sustainability. A defining feature of wetlands is that they are wet, at least for sufficient time for the wetness to determine the system's physical and biological attributes. Alterations to the hydrology of wetlands (including drainage, the regulation of timing and the volume of flooding events, changes to flow rates, water extraction, and chemical and temperature changes to water quality) will affect species composition and ecological processes. The drainage and manipulation of flooding regimes to permit intensification of agricultural use has a long history and has affected large areas of wetlands. "Reclamation" to eliminate wetlands through a combination of draining and filling, permitting a range of uses, has also affected many wetlands. In the twentieth century, a common scenario was the use of wetlands for the disposal of municipal waste, often followed by conversion to sports fields. Pollution has caused degradation of wetlands, both at catchment (and larger) scales and at specific sites. At large scales, eutrophication, in which the wetlands are exposed to excess nutrients from agricultural and urban runoff, as well as inputs of agricultural and industrial chemicals, have affected many wetlands. At some sites, point-source discharges and the effects of major accidents such as oil spills have been devastating.

A major threat to biodiversity is the spread of introduced species, which we can regard as a form of biological pollution. Wetlands are not immune from invasion by non-native species, and they are affected by some of the world's worst environmental weeds. One of the world's worst weeds is the water hyacinth *Eichhornia crassipes*, which carpets waterbodies blocking irrigation and drainage canals, and on dying, creates anaerobic condition that can cause mass death of fauna.

Future Threats

Despite the widespread acceptance by governments (over 150 countries have signed the Ramsar Convention) and the broader community of the importance of promoting the sustainable use of wetlands, threats continue. The rapidly growing human population has a great requirement for freshwater, and without stringent measures it is likely that the demand for water will exceed what can be supplied sustainably. The capacity of water regulators to provide environmental flows to sustain wetlands will be limited, and the continuing urbanization of the global population will place pressure on available land, which is likely to be irresistible in the coastal zone. The massive reclamation at Saemangeum in South Korea in 2006 has displaced many local people who previously depended on harvesting shellfish from the intertidal mudflats, and it has destroyed one of the most important habitats for migratory shorebirds in northeast Asia. Even with the global protests from environmentalists, it is unlikely that Saemangeum will be the last example of large-scale wetland loss. The increased human population will place pressure on global food resources, in turn generating pressure for increased access to water and land. Intensified agricultural use of wetlands, diversion of water for irrigation, and drainage for new agriculture are likely to continue. Especially in the developing world, pollution will remain a major issue, while the spread of introduced species shows little sign of abatement.

A major concern of many people is the consequence of increasing atmospheric greenhouse gases. Increases in carbon dioxide concentration will alter the relative abundance of species and wetland productivity. It is also extremely likely that increased greenhouse gases will result in climate change. Increased temperature will affect the abundance and distribution of species, while changes in rainfall and storm regimes could have major impacts on many wetlands. Of particular relevance for coastal wetlands will be the consequences of sea level rise. In many locations, it is probable that "coastal squeeze" will occur, as artificial barriers (embankments, housing, industrial or infrastructure developments) or natural topography prevent coastal communities from retreating inland. Nevertheless, landward retreat will be possible on some extensive coasts (e.g., in the Arctic and tropical Australia) and in other regions where wetland growth will be able to compete with sea level rise. Sea level has changed over geologic time, and communities have been able to migrate, but the alienation of so much of the coastline by human development means that nature will no longer always be able to look after itself.

One reason for modifying wetlands was to minimize disease risks to humans. While we long ago rejected the concept of miasmas—gaseous emissions from wetlands—as being the cause of disease, the role of the insects associated with wetlands as vectors of human disease is better founded. With human population increasing within the insect range of wetlands and with climate change affecting the incidence of both insects and disease agents, there is likely to be increasing pressure to either drain or spray wetlands with insecticides for health reasons.

Restoration, Rehabilitation, and Creation

As well as the upsurge in active management of wetlands, there is considerable activity in restoring and rehabilitating wetlands that were damaged by past human activity. In many instances, governments have mandated these works as compensation for their approval of new developments.

Despite problems with setting and implementing objectives for restoration and a failure in many cases to assess outcomes, there is no doubt that as more knowledge accumulates, we are better able to plan future projects. Most of the very large number of projects around the world have been single, often small sites, but large landscape-scale projects are also being contemplated, for example, in the Florida Everglades and the Tigris-Euphrates delta on the Iran-Iraq border, the home of the Marsh Arabs and their unique culture (Thesiger 1964; Maxwell 1957; Zedler 2006).

Humans have a long history of creating artificial wetlands, either deliberately in the landscape gardening tradition or accidentally (as, for example, following gravel extraction or as a result of mining subsidence). Some of what we regard as important wetlands are of artificial origin, such as the Norfolk Broads in eastern England, which were the result of medieval peat mining. There is

global interest in constructing artificial wetlands for wastewater management. These systems can remove nutrients from gray water, and their scale ranges from miniature reed swamps a few meters square, servicing a single dwelling, to larger systems serving neighborhoods or industrial plants. Wetlands have been installed on the roofs of multistory apartment complexes. Not all planning authorities have embraced this use of wetlands, but success has been achieved in a wide variety of conditions, suggesting that, in the future, wetlands will be increasingly important in developing sustainable suburbs. While these artificial wetlands will be used for nutrient reduction, and they will need to be managed as such, they will also have aesthetic and habitat values contributing to the attractiveness of neighborhoods. Increasing concern about the potential for these systems to support mosquitoes and thus pose a risk to the human population can be addressed with appropriate management.

Paul ADAM
University of New South Wales

See also in the *Berkshire Encyclopedia of Sustainability* Agriculture (*several articles*); Conservation Value; Fertilizers; Greenbelts; Insects—Beneficial; Insects—Pests; Nitrogen; Rivers; Soil; Water (Overview); Water Energy; Wise Use Movement

FURTHER READING

Batzer, Darold P., & Sharitz, Rebecca R. (Eds.). (2006). *Ecology of freshwater and estuarine wetlands.* Berkeley: University of California Press.

Callaway, John C., & Zedler, Joy B. (2004). Restoration of urban salt marshes: Lessons from Southern California. *Urban Ecosystems, 7,* 107–124.

George, Martin. (1992). *The land use, ecology and conservation of Broadland.* Chichester, UK: Packard Press.

Giblett, Rodney J. (1996). *Postmodern wetlands: Culture, history, ecology.* Edinburgh, UK: Edinburgh University Press.

Hey, Donald L., & Philippi, Nancy S. (1999). *A case for wetland restoration.* New York: John Wiley and Sons.

Humphreys, William F. (2006). Aquifers: The ultimate groundwater-dependent ecosystems. *Australian Journal of Botany, 54,* 115–132.

Hunt, Janet. (2007). *Wetlands of New Zealand: A bitter-sweet story.* Glenfield, New Zealand: Random House.

Kingsford, Richard. (Ed.) (2006). *Ecology of desert rivers.* Cambridge, UK: Cambridge University Press.

Marshall, Curtis H.; Pielke, Roger A.; & Steyaert, L. T. (2003). Wetlands: Crop freezes and land-use change in Florida. *Nature, 426,* 29–30.

Maxwell, Gavin. (1957). *A reed shaken by the wind.* London: Longmans.

Mitsch, William J., & Gosselink, James G. (2007). *Wetland ecosystems.* Hoboken, NJ: John Wiley and Sons.

Mitsch, William J.; Gosselink, James G.; Anderson, Christopher J.; & Zhang, Li. (2009). *Wetland ecosystems.* Hoboken, NJ: John Wiley and Sons.

National Research Council. (2001). *Compensating for wetland losses under the Clean Water Act.* Washington, DC: National Academy Press.

National Wetlands Working Group. (1988). *Wetlands of Canada.* Ottawa, Ontario: Environment Canada.

Oxford English Dictionary. (2010). OED Online. Retrieved October 12, 2011, from http://www.oed.com/view/Entry/161266?redirectedFrom=regime

Perillo, Gerardo; Wolanski, Eric; Cahoon, Donald R.; & Brinson. Mark M. (Eds.). (2009). *Coastal wetlands: An integrated ecosystem approach* (1st ed.). Amsterdam: Elsevier.

Polunin, Nicholas V. C. (Ed.). (2008). *Aquatic ecosystems: Trends and global prospects.* Cambridge, UK: Cambridge University Press.

Ramsar Convention. (1971). Convention on Wetlands of International Importance especially as Waterfowl Habitat. Ramsar (Iran), 2 February 1971. UN Treaty Series No. 14583. As amended by the Paris Protocol, 3 December 1982, and Regina Amendments, 28 May 1987.

Rieley, J. O., & Page, S. E. (Eds.). (1997). *Biodiversity, environmental importance and sustainability of tropical peat and peatlands.* Tresaith, UK: Samara Press.

Shine, Clare, & de Klemm, Cyrille. (1999). *Wetlands, water and the law: Using law to advance wetlands conservation and wise use.* Gland, Switzerland: IUCN Environmental Law Centre.

Silliman, Brian R.; Grosholz, Edwin D.; & Bertness, Mark D. (Eds.). (2009). *Human impacts on salt marshes: A global perspective.* Berkeley: University of California Press.

Teal, Mildred, & Teal, John. (1964). *Portrait of an island.* New York: Atheneum.

Thesiger, Wilfred. (1964). *The marsh Arabs.* London: Longmans.

Vileisis, Ann. (1997). *Discovering the unknown landscape: A history of America's wetlands.* Washington, DC: Island Press.

Williams, Michael. (Ed.). (1991). *Wetlands: A threatened landscape.* Oxford, UK: B. Blackwell.

Willott, E. (2004). Restoring nature, without mosquitoes? *Restoration Ecology, 12,* 147–153.

Woodley, Keith. (2009). *Godwits: Long-haul champions.* North Shore, New Zealand: Raupo.

Zedler, Joy B. (2001). *Handbook for restoring tidal wetlands.* Boca Raton, FL: CRC Press.

Zedler, Joy B. (2006). Wetland restoration. In Darold P. Batzer & Rebecca R. Sharitz (Eds.), *Ecology of freshwater and estuarine wetlands* (pp. 348–406). Berkeley: University of California Press.

Index

A

acid rain, 101
Acid Rain Emissions Trading Program, 31
Adirondack Park, 119
aerosols, 8, 65, 66, 67
Africa, 14, 17, 50, 61, 71, 74, 80
   buffers in, 26
agro-ecosystems, 95
agroforestry, 4
aircraft, 80
air pollution, 7, 28
Air Pollution Control Act of 1955 (United States), 28
Air Pollution Indicators and Monitoring, 7–12
Air Quality Act of 1967 (United States), 28
air quality control regions (AQCR), 29
Alaska, USA, 61, 63, 133
algae blooms, 122
Alps, 63, 72, 130
the Amazon (Amazonia), 3, 97
   rain forests in, 101
Amazonia, 60. *See also* Tropical rainforests
Amazon River, 13. *See also* **Rivers**
American Lung Association *State of the Air 2010* report, 31
American Society of Landscape Architects, 84
Andes, 61–63, 120, 128
animal husbandry, 74
Antarctica, 14, 61–63, 70, 77

anthropogenic causes and/or effects
   climate change, 64, 66–67
   nitrogen saturation and cycling, 103–104
Appalachian Trail, 120
aquaculture, 48–50, 52, 54. *See also* **Fish**
Aquifers, 13–21, 63, 128, 130
Aquifer Storage and Recovery (ASR), 19
Arctic, 52, 62
Argentina, 14, 62
arsenic, 129
artesian wells, 15
ARtificial Intelligence for Ecosystem Services
   (ARIES), 3
Asia
   climate change, 65
   greenhouse gases, 67
aspirin, 86
Atlantic Ocean, 49–51, 53–54
   coastal systems management in, 34
   garbage patches in, 106
   warming in, 66
   western Atlantic, predator fi sh levels in, 57
Australia, 14, 17, 74, 79–81, 119–121, 128
   agriculture in, 113
   climate change, 65
   ecosystem services, 3
   forests of, 139
Australian Qualifications Framework, 113

Bold entries and page numbers denote article titles in this book.

Bold entries and page numbers denote article titles in this book.

Bold entries and page numbers denote article titles in this book.

Bold entries and page numbers denote article titles in this book.

Image Credits

The illustrations used in this book come from many sources. There are photographs provided by Berkshire Publishing's staff and friends, by authors, and from archival sources. All known sources and copyright holders have been credited.

Bottom front cover photo is of fireflies (*Pyractomena borealis*) on an Iowa prairie, by Carl Kurtz.

Engraving illustrations of plants and insects by Maria Sibylla Merian (1647–1717).

Beetle, dragonfly, moth, and ladybug illustrations by Lydia Umney.

Front cover images, left-to-right:

1. *Tibetan glaciers.* Photo by Dan Miller, USAID.
2. *Bees on thistle, Highland County, Virginia, USA.* Photo by Ellie Johnston.
3. *Stormy sky and cows.* Photo by Amy Siever.

Back cover images, left-to-right:

1. *Queen angelfish, Flower Garden Banks National Marine Sanctuary, Gulf of Mexico.* Photo by Joyce and Frank Burek. National Oceanic & Atmospheric Administration (NOAA).
2. *Iguana in a tree, Belize.* Photo by Liz Wilkes.
3. *Mushrooms.* Photo by Amy Siever.

Pages VII, VIII, 156, and 167, *Water lilies.* Photo by Carl Kurtz.

Page 1, *Water lily.* Photo by Jill Jacoby.

Page 7, *Blue Ridge Mountains. Virginia. USA.* Photo by Ellie Johnston.

Page 13, *Connestee Falls, Brevard, North Carolina, USA.* Photo by Ellie Johnston.

Page 22, *Rose mallow.* Photo by Carl Kurtz.

Page 24, *American bison, Yellowstone National Park, Wyoming, USA.* Photo by Amy Siever.

Page 28, *Sunset, Hawaii, USA.* Photo by Zhang Xiaojin.

Pages 33 and 105, *Coastline, Cape Town, South Africa.* Photo by Ellie Johnston.

Pages 40 and 45, *Rudbeckia 'Prairie Sun' flowers, Butchart Gardens. British Columbia, Canada.* Photo by Ellie Johnston.

Page 48, *Fish at the Georgia Aquarium, Atlanta, Georgia, USA.* Photo by Ellie Johnston.

Page 55, *Partially burned forest near Mt. Humphreys, northern Arizona, USA.* Photo by Bill Siever.

Page 59, *Prairie fire.* Photo by Carl Kurtz.

Pages 61 and 119, *Tibetan glaciers.* Photo by Dan Miller, USAID.

Page 64, *View from Olana looking south, Hudson Valley, New York, USA.* Photo by Larry Lederman.

Page 70, *Giraffe, Kenya.* Photo by Kelly and Robin. Courtesy of morguefile.com.

Page 76, *Hot pool, Yellowstone National Park, Wyoming, USA.* Photo by Amy Siever.

Page 78, *Tropical plant, Manu National Park, Peru.* Photo by Ellie Johnston.

Page 82, *Night sky, Bruce Peninsula, Georgian Bay, Ontario, Canada.* Photo by Robert Dick.

Page 86, *Fossil shells embedded in rock at Castle Point, New Zealand.* Courtesy of morguefile.com.

Pages 89 and 94, *Agapanthus, Cape Town, South Africa.* Photo by Ellie Johnston.

Pages 96 and 100, *Cows drinking, Naumkeag Reservation, Stockbridge, Massachusetts, USA.* Photo by Amy Siever.

Pages 109 and 115, *Goats in an olive grove, Chaffin Family Orchards, Oroville, California, USA.* Photo by Chris Kerston.

Page 122: *View of US Route 180 and Mt. Humphreys, northern Arizona, USA.* Photo by Bill Siever.

Page 127: *Aerial photo of the Indus River Valley.* Photo courtesy of NASA.

Page 131: *Garden, France.* Photo by Ellie Johnston.

Page 136: *Arabian oryx, Phoenix Zoo, Phoenix, Arizona, USA.* Photo by Janet Tropp.

Page 141: *General view of the city and the Atchison, Topeka, and Santa Fe Railroad, Amarillo, Texas; Santa Fe R.R. trip.* Photo by Jack Delano, Library of Congress.

Pages 145 and 150: *Ruby Beach sunset, Washington, USA.* Photo by Ellie Johnston.

Author Credits

Introduction: What Are Ecosystem Services?
by **Amy Rosenthal, Emily McKenzie,** and
Kimberly Lyon
World Wildlife Fund

Air Pollution Indicators and Monitoring
by **Ian Colbeck** and **Zaheer Ahmad Nasir**
University of Essex

Aquifers
by **Glenn Longley** and **Rene Allen Barker**
Texas State University

Biocentrism
by **Gavin Van Horn**
Southwestern University

Buffers
by **Sarah Taylor Lovell**
University of Illinois, Urbana–Champaign

Clean Air Act
by **John Copeland Nagle**
Notre Dame Law School

Coastal Management
by **Richard Burroughs**
University of Rhode Island

Ecosystem Health Indicators
by **David J. Rapport**
EcoHealth Consulting and **Mikael Hildén**
Finnish Environment Institute (SYKE)

Externality Valuation
by **Areti Kontogianni**
University of the Aegean

Fish
by **D. G. Webster**
Dartmouth College

Food Webs
by **Jennie R. B. Miller** and **Oswald J. Schmitz**
Yale School of Forestry & Environmental Studies

Forests
by **Elizabeth A. Allison**
University of California, Berkeley

Glaciers
by **Mark Carey**
University of Oregon

Global Climate Change
by **Charles E. Flower, Douglas J. Lynch,** and **Miquel A.
Gonzalez-Meler**
University of Illinois at Chicago

Grasslands
by **Lynn Huntsinger**
University of California, Berkeley
and **Li Wenjun**
Peking University

Greenhouse Gases
by **John G. Stevens**
University of North Carolina Asheville

Insects
by **Richard A. Jones**
Royal Entomological Society

Landscape Architecture
by **Frederick Steiner**
University of Texas at Austin

Medicinal Plants
by **Lyle E. Craker**
University of Massachusetts

Microbial Ecosystem Processes
by **Jessica L. M. Gutknecht**
Helmholtz Centre for Environmental Research—UFZ

Mutualism
by **Ginny M. Fitzpatrick** and **Judith L. Bronstein**
University of Arizona

Natural Capital
by **Joshua Farley**
University of Vermont

Nutrient and Biogeochemical Cycling
by **Deane Wang**
University of Vermont

Ocean Resource Management
by **Kateryna M. Wowk**
Global Ocean Forum

Permaculture
by **Ian R. Lillington**
Swinburne University, Australia

Plant–Animal Interactions
by **T. Michael Anderson**
Wake Forest University

Recreation, Outdoor
by **Laura Anderson** and **Robert Manning**
University of Vermont

Resilience
by **Craig R. Allen**
US Geological Survey, Nebraska Cooperative Fish & Wildlife Research Unit, University of Nebraska
Ahjond S. Garmestani,
US Environmental Protection Agency, National Risk Management Research Laboratory
and **Shana M. Sundstrom**
Nebraska Cooperative Fish & Wildlife Research Unit, University of Nebraska

Rivers
by **Mark Cioc**
University of California, Santa Cruz

Soil
by **Timothy Beach**
Georgetown University

Succession
by **Lynn Huntsinger** and **James W. Bartolome**
University of California, Berkeley

True Cost Economics
by **William E. Rees**
University of British Columbia

Water
by **Brent M. Haddad**
University of California, Santa Cruz

Wetlands
by **Paul Adam**
University of New South Wales

This **BERKSHIRE** *Essentials* book was distilled from the

Berkshire Encyclopedia of Sustainability VOLUMES 1–10

Knowledge to Transform Our Common Future

In the 10-volume *Berkshire Encyclopedia of Sustainability*, experts around the world provide authoritative coverage of the growing body of knowledge about ways to restore the planet. Focused on solutions, this interdisciplinary print and online publication draws from the natural, physical, and social sciences—geophysics, engineering, and resource management, to name a few—and from philosophy and religion. The result is a unified, organized, and peer-reviewed resource on sustainability that connects academic research to real world challenges and provides a balanced, trustworthy perspective on global environmental challenges in the 21st century.

Ray C. Anderson

General Editor

Sara G. Beavis, Klaus Bosselmann, Robin Kundis Craig, Michael L. Dougherty, Daniel S. Fogel, Sarah E. Fredericks, Tirso Gonzales, Willis Jenkins, Louis Kotzé, Chris Laszlo, Jingjing Liu, Stephen Morse, John Copeland Nagle, Bruce Pardy, Sony Pellissery, J.B. Ruhl, Oswald J. Schmitz, Lei Shen, William K. Smith, Ian Spellerberg, Shirley Thompson, Daniel E. Vasey, Gernot Wagner, Peter J. Whitehouse

Editors

10 VOLUMES • 978-1-933782-01-0
Price: US$1500 • 6,084 pages • 8½ × 11"

"This is undoubtedly the most important and readable reference on sustainability of our time"

—Jim MacNeill, Secretary-General of the Brundtland Commission and chief architect and lead author of *Our Common Future* (1984–1987)

"The call we made in *Our Common Future*, back in 1987, is even more relevant today. Having a coherent resource like the *Encyclopedia of Sustainability*, written by experts yet addressed to students and general readers, is a vital step, because it will support education, enable productive debate, and encourage informed public participation as we join, again and again, in the effort to transform our common future."

—Gro Harlem Brundtland, chair of the World Commission on Environment and Development and three-time prime minister of Norway

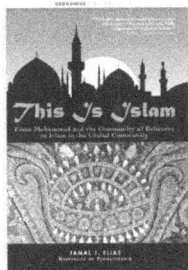

www.ingramcontent.com/pod-product-compliance
Lightning Source LLC
Chambersburg PA
CBHW080423270326
41929CB00018B/3141